Confessions of a Criminal Cockroach

The Ultimate Cockroach Combat Manual

Second Editon

Guy Amato

ISBN: 979-8-950722-00-4

First Edition: 2025

Second Edition: 2026

Story Harbor Publishing

Printed in the United States of America

CONTENTS

PREFACE

It's the same old story. I have heard it more times than I can count. Someone sees a cockroach in their kitchen ... most likely at night or very early in the morning, when it is still dark outside. It happens so fast that you may wonder if you even saw it at all. Something small scurries away at a lightning pace. You try to convince yourself that you were just seeing things, or perhaps it was just a fly or something unimportant, so you go back to bed, or go on getting ready for the day. This may happen more than once before you realize that what you saw was a cockroach.

Sure, you're a little freaked out at this point, but not knowing much about cockroaches, you still figure it's probably no big deal. Then one day you start seeing them during the day. At that point, you may conclude that you have a "minor" infestation of cockroaches.

You then set about the task of trying to get rid of them. You tried putting out sticky traps, and while you did catch some roaches, they kept coming. You may have moved on to some pesticides that you found at your local store that were labeled for cockroaches. It didn't take long before you realized that this wasn't working either. Perhaps you conducted a little research and found recommendations that you should use a bait. So, you gave that a shot as well. But again, the roaches just kept coming. In fact, the infestation started getting worse than ever. So, in desperation, you hired a pest controller.

You are now totally confident that the problem is solved because surely a professional pest controller knows exactly what to do to end the infestation. Sure enough, it didn't take long before you stopped seeing any cockroaches. Sometimes the pest controller will just do a one-time treatment, but more often than not, you will be sold an ongoing treatment plan. Either way, the treatment worked for a while, then gradually, the roaches came back. In fact, they were back worse than ever. Now, out of total exasperation and a feeling of utter defeat, you fired the pest controller and decided to just live with the roaches. It never ceases to amaze me how many people end up in this situation.

You may also be a restaurant owner who is busy trying to keep the health department from writing violations because of cockroaches, so you hire a pest controller, who visits routinely to fog your entire restaurant. You get a nice body count, but the problem is always ongoing. The roaches just keep coming and they never seem to stop.

It is no different if you are in charge of running a hospital, hotel, apartment building, or any other type of business establishment that has a cockroach infestation. Almost everyone concludes that you just can't get rid of these things. The best you can hope for is to try and keep them under control.

You may even be a pest controller yourself, and maybe you are wondering if there's any possible way to offer a treatment that will completely eliminate German cockroaches.

If you are in any of these situations, then I have good news for you. There is a way to totally eliminate German cockroaches from your home, business, and even your car. Contained within the pages of this book is the solution to your problem. I'm not going to lie and tell you that it's easy, but I have done this myself, multiple times, and it does work. In fact, if you follow the directions very carefully, this works every single time. I have shared this treatment plan with over half a million people, and not one of them has reported that it didn't work.

I promise … this is like nothing you've tried before. The best part is that almost anybody can do it. So, suffer no more with roaches. You're holding the solution to the problem in your hands. All you need to do now is implement it, and your nightmare is over.

WHO IS RONNIE THE ROACH

Being involved in the pest control industry for over 40 years has required me to read many pest control books, studies, articles, and other written material. Even though pest control was my job, I still found all of these to be somewhat laborious to read. I decided that, if I must write a pest control book, then I needed a way to make it a bit more interesting for folks who know little or nothing about pest control. Ronnie popped into my imagination one evening without warning. The thought occurred to me that perhaps this material might be somewhat less boring to read if it was told by a roach. There was no specific reason for naming him Ronnie, other than, well, it just seemed to work. Besides, my wife suggested the name, which, of course, is more than reason enough to use it.

For those of you who may be interested, I want you to know that, even in the darkest corner of my imagination, I never considered actually spraying Ronnie with a pesticide. The truth is that I was just bluffing. You may be pleased to learn that I imagine him living out the rest of his life, as brief as it may be, in a habitat designed for total cockroach comfort and luxury.

Even though Ronnie was a big help with the writing of this book, I think it's safe to say that he will remain retired, which I think is completely appropriate in keeping with the fact that I have no intention of writing another book. At least not about German cockroaches. Of course, I have learned over these many years to never say never in pest control, so who knows?

1 I'M RONNIE THE ROACH

Hi, my name is Ronnie. Now, I don't want you to freak out or anything, but you should know that I am a roach. Not just any roach mind you. I am a Blattella germanica. More commonly known as a German cockroach. If you are familiar with my species, then you probably think that I am a vile and disgusting creature with no business occupying any space on the planet whatsoever. Well, I hate to admit it, but the truth is that you are right. Here's the thing though … my species has been around for millions of years. We have seen the dinosaurs come and go, we survived the Ice Age, and we even survived the atomic blasts on Hiroshima and Nagasaki. So, you might say, I am something of a survivor.

Now, you might be wondering why I am speaking with you today, and if you are not pondering that question, then you should be. The answer is simple. I am here to explain the only way that you can get rid of me, and the rest of my species, once and for all, forever, out of your life. Also, for those of you who do not have an existing infestation of German cockroaches already, I will explain how to prevent one.

If you are somebody who has tried everything you can possibly think of to get rid of us, and you've had no success, there's a good reason for your failure. It's pretty simple really. The problem is that you do not understand how we think and behave, and if you don't understand how we think and behave, you are never going to be successful in totally eliminating us from your environment. So today, I am going to explain all of this and provide the secret about how you can eliminate German cockroaches from your life for good.

You may also be wondering why in the world a German cockroach would be giving you the secret about how to eliminate his own kind. Well, like I said … I am a survivor. As repugnant as this may sound to a human with any sort of moral compass, I have been left with little choice but to reveal these secrets. This was not a hard decision for me because, unlike you humans, I have no moral compass whatsoever.

You see, I was recently captured by a clever pest controller named "Guy," and he coerced me into spilling the beans so to speak. More to the point, he had me in a jar and he was going to spray me with pesticide if I didn't confess to my crimes of spreading disease and delighting in the ridiculous antics of humans who think they can defeat us. Furthermore, he insisted that I explain, in meticulous detail, how to eliminate my species from the face of the Earth. He's kind of mean if you ask me, but I can't help but be impressed by how well he has learned to speak German cockroach. I have never encountered a human who could understand what we were saying. It's like he knows us so well; he could be one of us.

So, there you have it. I sold out the entire German cockroach species to save my own exoskeleton. Again, no moral compass here. German roaches are not exactly known for carrying out acts of heroism, and I am no exception.

Anyway, enough about me. Let's get right into the information you need to obliterate every German cockroach in your home, business, or car.

2 DENIAL - KNOW THE SPECIES

You know, one of the big differences between you humans and us is that, when we encounter a threat, we run. We run every single time. There is no fight or flight response with us. With you humans, you will actually take the time to think about fighting your enemies. That makes no sense, but what's worse is that many times, you will be so afraid that you may have a serious problem that you will ignore it. A good example of this may be that you are feeling some sort of pain in your abdomen. While it would make good sense to get checked out by a doctor, you may fail to do that, because you are afraid of what he or she is going to tell you. Similarly, if you go out into the kitchen in the middle of the night and turn on the light, you may see one or two of us, yet you do nothing. Sure, some of you will go into that fight or flight mode, and you will either try to kill us, or run away like a frightened nymph. But mostly though, you may do what humans do so often. You go into your denial mode and try to tell yourself that you do not have an infestation of German cockroaches, or that it's only a "minor" infestation. After all, you only noticed one or two of us, so how bad can it be?

To make matters worse, this misconception may be reinforced if you manage to kill those one or two roaches. At that point, you think you solved the problem, and you go back to bed. Deep down though, you must know that, if you see one or two roaches, there's got to be others lurking about as well. However, you don't want to admit to yourself that you have a serious cockroach problem, so you convince yourself that you don't. It totally amazes me that humans run around bragging all the time about how big and beautiful their brains are, yet they can be incredibly unwise sometimes.

Well, before you go back to bed, let me give you a wake-up call. If you see even one German cockroach, then you have a serious infestation. A lot of people will see one or two of us at night, and sometimes during the day, and think that they have a minor infestation. No big deal. Nothing to worry about. Right? Good grief. It is absolutely mind-blowing how clueless you big-brained humans can be. HELLO! There is no such thing as a minor German cockroach infestation. If you see even one cockroach, then there are many, many more of us lurking in hiding places that you cannot see. When you start seeing us during the day, even if you are only seeing one or two of us, then I can assure you that there are hundreds, and maybe even thousands, of us in your environment. So, snap out of it! When you see a German cockroach in your home, then you need to get out of this denial mode and understand that you have a serious problem. If left unchecked, you will quickly get overrun by us within a couple of years.

If you do not know who we are, then you need to educate yourself. Part of your denial mode is that you will convince yourself that we are not German cockroaches at all, but rather some other sort of roach that is not going to be a problem. Look ... it is true that no other species of cockroaches are as bad as us. It is also true that some species, such as our Asian cockroach cousins, are unable to infest homes. They will fly to your house because they are attracted to light. One of them may make it in and then end up dying because they cannot survive in your environment. Since you don't know one cockroach from another, your ridiculous denial mode will kick in, and you will convince yourself that you have a species that is no problem. The truth is, if you see even one roach in your environment, then at a minimum, you need to identify the species. Assuming that you are looking at a harmless species, without any facts to back it up, is extraordinarily unwise and one of the worst mistakes you can possibly make.

So, let me give you a little tip. If you see even one roach in your environment, then obtain a specimen. A good clue that you may have German cockroaches is that we are very fast. So, if you cannot catch one of us, then get some sticky traps and place one under the kitchen sink, under the refrigerator, and under the stove. If you have German cockroaches, then you should catch at least one of us over the next couple of days. After you obtain a specimen, then compare it to the photos in Figures 1.1 and 1.2, to see if it's an adult or nymph. Nymphs are just babies or child cockroaches that have not grown up yet. You should be able to tell right off if it looks like me. If it doesn't look like me, then congratulations. You can stop reading. You don't need to do all these treatments. Instead, just go to http://www.guyspestsolutions.com for the solution to your problem. However, if you confirm that you do have German cockroaches, then you would be well advised to read this book very carefully.

Figure 1.1 German Cockroach Adult

Figure 1.2 German Cockroach Nymph (Baby roach)

If you have confirmed that you DO have a German cockroach infestation, the first thing you need to do is learn something about German cockroach biology and habits. I know it is incredibly boring, but without this understanding, you won't stand any chance of beating us. Trust me. We are just going to laugh at you, and you don't want that to happen. I mean, it's really embarrassing to be beaten and laughed at by a roach. I understand how tempting it is to kind of skip over how we think and behave, but if you do that, then you are apt to make some serious mistakes that are likely to end in failure.

These are the same mistakes that even professional pest controllers make. They think they know a lot about us, but the truth is, they know very little about us. If you do not understand how we operate, then the biggest mistake you are likely to make is that you will underestimate us. Take it from me ... if you want to fail, just underestimate us. We will leverage every mistake you make into a total disaster on your part. Don't worry though. I will keep the whole biology thing just as brief as possible.

3 GERMAN COCKROACH HABITS

The first thing you need to know about me is that I am a domestic roach. That is to say… I do not survive well outside. I only live inside buildings, so I am dependent on people. It kind of makes you wonder how we survived for millions of years before people were even on the planet. In the interest of brevity, I will save that story for another time. For now, let's just say, we evolved over the years and without you, there would be no me. So, if you are looking for someone to blame, just glance into the closest mirror.

While there are predators that would love to eat us, you seldom find them in homes or businesses. Even if you do, we multiply way too fast to make introducing predators an effective treatment plan. We can multiply so fast because we only have two functions in life. They are to survive and reproduce. Believe me … we are very good at both of those things. In fact, we are probably better at surviving and reproducing than almost any other living organism on the face of the planet. These abilities are enhanced further by that predictable, silly human behavior where people always underestimate our capabilities. You folks just make our survival way too easy.

Take food and water for example. You see, contrary to what most people believe, we do not need to come out of hiding to get food and water. Sounds crazy. Right? We all know that all living things need food and water to survive, and that is true about German cockroaches too. The thing is though that we can eat all sorts of materials that can be found in your wall voids and other hiding areas. Best of all, we have the ability to survive by eating each other. That's right. We are cannibalistic. When one of us dies, all we need to do is eat the dead roach. We do not eat very much, so one dead roach can feed a whole bunch of other roaches for some time. In fact, not only can we survive like that, but we can multiply in mass numbers.

Just one German cockroach egg sac will produce between 30 to 40 nymphs (baby roaches). I'm sure that any human woman who has experienced childbirth will probably empathize with our females. Before you feel too bad for those lady roaches though, keep in mind that they only lay egg sacs and not fully developed babies, so you human women have it much worse. We have a lifespan that can vary quite a bit, but since we have a life cycle of about 100 days, that means there is a steady supply of dead roaches for us to eat. This ability allows us to reproduce at such a rapid pace that our populations are going to expand dramatically, in a very short period, even without access to food and water. So, one of the biggest mistakes that people make is to think that they can simply eliminate our food and water sources and that will get rid of us. The truth is all we do is laugh at those people.

You can have the cleanest house in the world, but we don't care. We will infest any building that we can gain access to, regardless of how clean it is. Now, wait just a minute you say, "This cannot possibly be true." Everyone says to keep a clean house to help rid yourself of roaches. Hey… indeed, we would much rather eat some of your left-over eggplant parmesan, rather than a nasty cockroach, but the point is that we will go on breeding regardless of how clean you are. Keeping your home squeaky clean will discourage us, but it will not stop us.

If you are wondering how we can infest so many buildings, even though we do not survive well outside, the answer is pretty simple. We are hitchhikers. We live in restaurants, hotels, hospitals, stores, warehouses, and of course, people's homes. So, all you need to do for us to infest your home is go to a restaurant, hotel, or any place that happens to have an infestation. We will get into any personal belongings that you may have with you, or we may even climb on your clothes as well. Don't even think about spending the night in a hotel or somebody's house that is infested, because that is almost a sure recipe for bringing us back home with you. All it takes is for one female roach or egg sac to get into your belongings and you are going to have a massive infestation of German cockroaches.

How massive? Well, just do the math. An egg sac, or ootheca, contains between 30 to 40 eggs. If one female produces just 30 cockroaches every 100 days, and half of those roaches are females, then that is 15 females after the first 100 days. If each of those 15 females produces 30 offspring and half of them are females, now you have 225 female cockroaches, each producing another 15 females. If you run the math for a couple of years, even accounting for the roaches that die off, you will find that you have well over a million cockroaches in less than two years. Now, can you imagine what a million cockroaches will turn into two years after that? Also, consider that a female nymph (baby) may mature faster than 100 days. If conditions are right, a female can produce an egg sac after only 40 days. Also, even though our average life span is only about 100 days, we can survive for up to seven months under ideal conditions. Just like you, if we eat a healthy diet, get plenty of exercise, watch our weight, and see our primary care physician regularly, then we will live a lot longer. Just kidding. Ideal conditions for us are lots of food and water, accompanied by an environment that is climatically controlled. So, you cannot rule out the possibility that you may have those million roaches in less than a year.

If that isn't bad enough, another important fact for you to consider is that about 90% of us never come out of hiding. Like I said, we don't need to come out of hiding, because we can eat each other. Therefore, if you think you are going to eliminate a German cockroach infestation by putting out some sticky traps, bait, spraying some pesticide, and that sort of thing, then think again. We are just going to laugh at you. Sure, you may kill a lot of roaches, but for every roach you kill, there are many more of us that will survive. To make matters worse, we will become resistant to the pesticides that you use, and so will our offspring. By using baits and sprays, you are simply making us stronger. In the end, you are just creating an army of super roaches. This, my human adversaries, is why we are so hard to eliminate. So, if this is the case, you may be questioning how in the world is it possible to effectively eliminate a German cockroach infestation from your home, business, or car. Well, read on. Even though this has been the best-kept secret for thousands of years, I will explain exactly how to do it.

4 CLEANING AND SANITATION

I know, you just finished reading that I can invade your home and reproduce in mass numbers, even if you keep it squeaky clean. While all that is very true, that doesn't mean that you need to make it easy for us. The fact is that cockroaches do not taste all that good. Take it from me. I have eaten more than my share of roaches. I think Uncle Fred tasted the worst. He may have gotten into some of that bait you humans use before he died. Anyway, if you give me a choice between eating some leftover food that you leave on dirty dishes in the sink, or eating my relatives, I will probably opt for eating the food on the dirty dishes in the sink. Not only does your food taste better, but we may live a lot longer if we have better dietary choices than other roaches. Most roaches don't see it this way, but enough of us do to make cleaning worthwhile. Also, even though I can get the fluids I need from the bodies of dead roaches, that doesn't mean I don't enjoy a nice drink of water. Again, a water supply can help to extend my life. So, if you want to make life a little bit harder for us, and reduce our chances of reproducing quite as fast, then always make sure that everything is squeaky clean.

Never leave dirty dishes in the sink overnight. Wipe down the sinks before you go to bed and always pick up pet bowls before you go to bed. Never leave pet food or water down all night, or I will make a feast out of it. Your pets are my best friends. You also need to clean the burners around the stove, crumbs out of the toaster, and remove any other sorts of grease or other possible sources of food. You should also have a rule that you never, ever eat in any room except the kitchen and the dining rooms. This is particularly true if you have children. Heck, I like kids almost as much as pets. Those little rug rats are notoriously messy. When you drop crumbs behind the couch or in the couch cushions, then I would be more than happy to make a meal out of them. So, not only will you have us in your kitchen, but you will now create an infestation in your couch. How disgusting is that? Also, if you bring food into the bedrooms, I would be more than happy to take up residence there as well. While you may find it somewhat repugnant to spend the night in bed snuggling with me, I am more than happy to become a frequent bedfellow for you. So, making sure that we have no food sources, other than eating each other, will be a good start toward getting rid of us. To sum up… Even though cleanliness will not stop me, cleanliness is clearly my enemy. We will survive anyway, but you do not need to make it easy for us.

5 PESTICIDE INFORMATION

Another big mistake people make when trying to eliminate a German cockroach infestation is they do it without understanding how pesticides work. While it is very tempting to just use any pesticide you want, thinking that they are all the same, this would be a huge mistake. There are different kinds of pesticides and if you select the wrong ones, then you're going to be doomed to failure. Talking about pesticides scares me to death, but let's take a minute to talk about them anyway. I promise that this will be well spent time.

There are two categories of pesticides that are very important to understand. Essentially, pesticides are either repellent or non-repellent. As the name implies, repellent pesticides repel insects. We have a way of identifying this material as being harmful to us, so we avoid it. There are also non-repellent pesticides. These pesticides are great for treating cockroaches because we do not know that we are being poisoned when we come in contact with this material. I will explain more about why this is important in just a bit.

Pesticides also differ in the time it takes for them to kill the target pest. There are instant-kill pesticides, fast-kill pesticides, and slow-kill pesticides. Instant-kill pesticides are usually repellent, while fast-kill and slow-kill pesticides are usually non-repellant. Some pesticides combine instant-kill chemicals with fast-kill chemicals to produce a product that will kill instantly, but also leave a long-lasting residual. What sicko human thought of that?

You should also know that pesticides come in different forms. They may be applied as liquids, dust, bait, fumes, or gas. An effective German cockroach elimination program must use at least a combination of liquid, dust, and bait. In certain cases, fumes can be used for limited applications. A big mistake that even the pros make is that they try to eliminate a German cockroach infestation by just using one of these forms, such as baits. This is never going to get the job done. Even though I am beginning to question the wisdom of this, I am going to show you how you can use a combination of these products to effectively reduce our population very quickly and finish us off in a way that will prevent a reoccurring infestation.

The most common way that pesticides kill is they work on our nervous system. Instant kill, nervous system pesticides typically utilize type 1 pyrethroids. This material results in a rapid knockdown. Roaches that come in contact with this material will become almost instantly paralyzed. Most amateurs love this because they get to see the roach flip over on its back with its legs uncontrollably kicking in the air. While you may get a satisfying and somewhat sadistic kick out of watching this, I can assure you that it is not an effective way to eliminate a German cockroach infestation. If you spray this material on a cockroach, then you just killed one cockroach. While that may feel like a victory, it is not going to achieve your goal. What you need to do is get those cockroaches working for you. A dead roach is not going to be able to carry any pesticide back to its hiding place, where it can share this material with other roaches.

This brings me to fast-kill pesticides. This material also works on our nervous system, but there is a delayed reaction. Unlike instant kill pesticides, you will not see us flip over on our backs with our legs kicking in the air. Instead, roaches that are exposed to this material will return to their hiding places, where they will experience a state of paralysis, and ultimately death, within a few hours after exposure. The other cockroaches will eat them, and many of those cockroaches are going to die as well. So, if you are in the business of killing cockroaches, then you want to kill as many as you can, as efficiently as you can. To do this, you must allow us to assist you with distributing the pesticides. You should always keep in mind that about 90% of cockroaches never come out of hiding. This is why it is very important to bring the fight to them. Just a word of caution though. Roaches will often become resistant to fast-kill pesticides rather quickly.

While some of the pesticides work on the nervous system, there are also very effective products that do not work this way.

Some pesticides kill mechanically. An example of an instant-kill pesticide that works mechanically would be to mix some dish soap with water. It sounds simple, but it works. If you spray a cockroach with this mixture, it will die rather quickly. This is because we have a network of tiny tubes called "tracheae" that run throughout our bodies. They open to the outside of our bodies through small holes called "spiracles." Air enters through the spiracles and travels through the tracheae to deliver oxygen directly to our tissues. If you spray us with soapy water, it will cover the spiracles, and we will suffocate to death. Of course, as I have previously mentioned, instant-kill pesticides are not a good idea in most cases.

There are also slow-kill pesticides that do a wonderful job. These would include things like boric acid, diatomaceous earth, and Nibor-D. When roaches walk across this material, it gets stuck to their feet. Since we cannot tolerate anything being stuck to our feet, we will ingest it to get it off. This material will then cut holes in our stomachs, and we will dehydrate to death over the next few days. Although these materials are very slow to kill, they offer the advantage that they are permanent, and we can never become resistant to them. As long as you do not remove them, they will go on killing us forever. This means that surfaces treated with these materials can no longer become infested.

There are also insect growth regulators (IGRs) that do not kill cockroaches at all. This material simply prevents us from reproducing. It's kind of like birth control for cockroaches. I have to say that, even though IGRs are not a good thing for us, using them has certainly made our females a lot more promiscuous. I suppose every cloud has a silver lining. Seriously? You didn't really believe that our females care about birth control … did you? Just the opposite is true. Our females love to have offspring, so they hate birth control of any kind. Admit it. I'll bet I did fool you there for a second. Anyway, as with all living things, if we cannot reproduce, then we cannot survive. The beauty of using IGRs is that we do not need to come in contact with this material for it to work. If you are using the IGR disks, you can simply place them in strategic locations, and they will release microscopic particles of the IGR, up to 75 square feet, and prevent us from reproducing. This may be a very slow process, but it is also very effective.

As discussed earlier, there are temporary pesticides and permanent pesticides. Temporary pesticides lose their effectiveness, or residual, after a certain period. This is usually somewhere between instantaneously to nine months. Permanent pesticides, as the name implies, are materials that virtually last forever. Clearly, using permanent pesticides is a great idea when treating a German cockroach infestation. But it means you must be patient because permanent pesticides kill very slowly. Typically, most non-repellent pesticides have a kill time of several hours, but permanent pesticides, which are also non-repellent, may have a kill time of as long as several days. Nothing like a super slow and agonizing death. Thanks a lot.

This is why many people get confused when they are using permanent pesticides. They see us walking right through it and it seems to have no effect whatsoever on us. Most amateurs like to see cockroaches contact the pesticide and immediately start dying. As satisfying as that may be, permanent pesticides do not work that way. Permanent pesticides do a great job, but they require patience. Don't worry. A roach that crosses this material has signed its own death warrant. So just take your satisfaction in knowing that the roach will be dead in a few days.

Before starting any pest treatment, be sure to wear the proper personal protective equipment. This usually includes pants, a long-sleeved shirt, eye protection, rubber gloves, and a respirator. Let's face it, your lungs were designed to breathe air and not pesticide. So, use some common sense. These materials are safe for children to play on after they dry, except for dust, but getting this stuff in your eyes, on your skin, or inhaling it may not be such a good idea. Your lungs were designed to breathe air and not chemicals. Even breathing air freshener is probably not a good idea. Let me quote directly from WebMD. "Not only do most air fresheners have chemicals such as formaldehyde, petroleum distillates, p-dichlorobenzene, and aerosol propellants, they can also catch fire." If you don't believe it, just pick up a can of air freshener and read the warning label on the back of the can. There is a can of air freshener in our bathroom warning that inhaling the product in concentrated form can be harmful or fatal. That's right … I said fatal, like in you can die from it. Even though it is highly unlikely that any of the pesticides I am recommending will kill you, breathing pesticides, or any other chemicals, is a bad idea. So, always use caution when handling any sort of pesticide, and wear the recommended personal protective equipment. Dust should only be used where it will not be disturbed by people, pets, or air currents.

It is always a good idea to test any pesticide on an inconspicuous area of the surfaces you are going to treat to ensure that they will not stain the surface.

When using any pesticide, you should always read and follow label directions. If the label directions differ from the recommendations in this book, you should always follow the label recommendations.

6 FOGGING

I would be totally remiss if I didn't mention a little bit about fogging. This is a process where the entire space being treated is fogged with an aerosol. Professionals do this with a fogging machine that breaks the liquid down into very small particles that float around in the air for some time. This is typically done by using an instant-kill pesticide, and professional pest controllers usually do this to treat restaurants and apartments. The advantage is that you get a very high body count, and it looks like the cockroach infestation has been eliminated. Unfortunately, nothing could be further from the truth. While many of us are going to die, even more of us are going to escape deeper into the walls, and we will most likely exit into adjoining living spaces or businesses.

Apartment landlords often do this when they have a vacant apartment that they need to rent. Fogging allows them to keep the cockroaches away for a couple of weeks, thereby fooling prospective tenants into thinking that the apartment is cockroach-free. Restaurant owners like it because they can fool health inspectors into thinking that they have few, if any roaches in their establishment. Unfortunately, this is only a temporary fix, because the roaches will return over the next few weeks. Not only will other tenants in adjoining units now be overrun with roaches, but unsuspecting tenants will realize after they move in that they have a serious cockroach problem. This is not to suggest that all landlords or restaurant owners do this, but it is very common. Also, there are almost no pest controllers who know how to perform this treatment because it was developed by Guy, and it is still unknown to most of the pest control industry. Therefore, this treatment method has not been available to landlords or restaurant owners until now. Hopefully, as this new treatment for German roaches becomes common knowledge in the industry, more landlords and restaurant owners will take advantage of it.

The bottom line is, with one exception, you should never fog any building to treat a German cockroach infestation. Not only will this drive us deeper into the walls, making it more difficult to treat the infestation, but it is also a very mean thing to do to your neighbors. You certainly would not want them to do that to you, so you should not do it to them either.

As for landlords that do it, I hope they will decide to do better going forward. Instead of trying to trick unsuspecting tenants into renting an infested apartment, they should be doing a proper treatment to fully eliminate the infestation. They kind of got a pass before, but now that they have the solution, there is no excuse for not doing the right thing. I'm sure some landlords will take advantage of this information, but I have no doubt that others will not. You see? There are some humans who have questionable morals as well.

The same holds true for restaurant owners who fog their establishments in an effort to trick health inspectors. They may get away with it, but those roaches will be back to contaminate the food that their customers are going to consume. Not only could their customers get sick, because roaches carry diseases, but their customers may also bring some of these roaches back with them and infest their living space. This is why restaurants that are found to have any German roaches whatsoever should be shut down. It is unconscionable to allow these cockroach distribution centers to continue operating. All it takes is for one female cockroach to attach itself to somebody's clothing, and that person will bring that cockroach back to their home and start an infestation. This is why we love to hang out under tabletops. People never look under the tables, and it gives us the perfect vantage point to simply drop onto an unexpecting victim. So, tell me … who's worse? Us, or the restaurant owner that participates in this unsavory behavior. Restaurant owners also got a pass before because fogging seemed like the only solution, but now they need to step up and totally eliminate the problem.

Now, there is one exception where fogging may be a good idea. If you have a vacant building that is totally overrun with roaches, then fogging may be a good way to quickly knock down the population. This will make the rest of the treatment much easier to perform. Since the building is unoccupied, this will not create a problem for other people.

7 HOW ROACHES BECOME RESISTANT TO PESTICIDES

If you've been paying attention, you may recall that I mentioned that roaches will become resistant to pesticides that work on their nervous system. This is why you can only use any given pesticide formula just one time, or you risk resistance. But how does that work?

The simple answer is that when a cockroach consumes a bait or walks across a pesticide that works on the roach's nervous system, they do not die immediately. They return to their hiding places, where they will die within a few hours. The other roaches are going to most likely eat them. Since the pesticide attacks our nervous system, the roaches that eat the dead roach are going to be consuming some of the pesticide. Most of the roaches that consume the dead roach will also get a lethal dose, but not all of them. Some of them will escape with their lives, but since they were exposed to some of this material, they will build up a resistance to it. Not only will they become resistant to it, but their offspring will too. Further, this resistance will be passed along to future generations. Assuming that the lifespan of a German cockroach is around 100 days, and since most of these pesticides have a 90-day residual, you can see how the roaches may become resistant at about the same time you will need to reapply the product.

This is why you can only use a particular formulation one time and then retire it forever. So, keep in mind that if you previously used a pesticide with the same active ingredient, then you cannot use it again. Also keep in mind that if you hired a pest controller, he may have used the same active ingredient, so the roaches you are treating may already be resistant to the products you are using.

8 PHASE 1 - THE QUICK KNOCKDOWN

If you have a German cockroach infestation, then I'm sure you noticed that most of us are living in your kitchen. That is because you have a lot of places in the kitchen for us to hide and breed. You may also have plenty of food sources for us as well, so many of us will choose not to eat each other. You may also find a lot of us hanging out in the bathrooms too. There may not be much food there, but there sure is plenty of water. Since there are so many of us living in the kitchen and bathrooms, it would be a good idea to do something to reduce our population in those areas just as quickly as you can. There is a procedure to do this that I like to call "Phase 1" of the treatment plan.

In Phase 1 you need to use some quick-kill products that should knock down our population quickly. However, before you apply these quick-kill products, you need to put down a base of slow-kill products that are going to last forever. Believe it or not, there are products on the market that we cannot become resistant to, and their killing properties last virtually forever. Of these, the most effective ones are Nibor-D and boric acid. If you like, you can use diatomaceous earth (DE) instead of boric acid for dusting. While you can use DE as a liquid application, I think Nibor-D is by far the better option. I will talk about using DE as a liquid application in Chapter 11.

Step-by-step procedure:

1. The first thing you need to do is mop all the floors with whatever detergent you like and vacuum all the carpeting.

2. Next, mop all your floors with Nibor-D, mixed at a rate of 2.5 scoops of product to a gallon (3.8 liters) of cold water. A 4-fluid ounce measuring scoop comes with the product. Mixing a detergent with Nibor-D is not recommended.

While it is not required, mixing an insect growth regulator (IGR) with Nibor-D will make it more effective. My favorite is NyGuard IGR. See Figure 2. You should mix it at a rate of 0.4 ounces (12 ml) of NyGuard to one gallon (3.8 liters) of mixed Nibor-D and water.

I recommend that you do not mix more Nibor-D than you will use within two hours. I know that the label may allow for more time than that, but I find it works best when it's fresh. Don't forget to apply Nibor-D under the stove and refrigerator.

Please note that Nibor-D can make some floors feel sticky and it can also make them look hazy or shiny, depending on the type of floor surface. Hazy or shiny floors are just part of the price for eliminating German roaches, but sticky floors need to be mitigated.

If your floors feel sticky, make sure that you used the correct concentration. Remember that you are mixing the Nibor-D at a rate of 2.5 scoops per gallon of water. If the dilution rate is correct, then you may need to remove the Nibor-D and reapply it only under furniture and next to walls. To remove the Nibor-D, simply mop the floors with any detergent. Reapply the Nibor-D under all furniture where you will not contact it with your feet. Then, create a 6-inch (15 cm) band on the floor around those items. Also, create a 6-inch (15 cm) band on the floor next to all the walls.

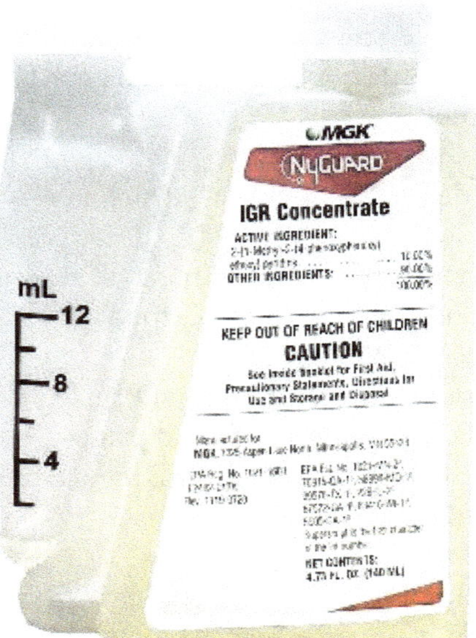

Figure 2 - NyGuard IGR

3. After mopping the floors with Nibor-D, it's a good idea to treat the carpets. This is not exactly consistent with label recommendations, so I cannot advise you to do this, but it will be a big help in eliminating the roaches. I personally do not think this treatment is harmful, but following these directions is a decision you must make for yourself. To me, it's just common sense that you need to prevent the roaches from moving freely across the carpets. Mix the Nibor-D at a rate of 2.5 scoops per gallon (3.8 liters) of cold water and lightly spray the carpet under upholstered furniture and create a 6-inch (15 cm) band on the floor around walls. You can apply the product with a pump-up garden sprayer or a handheld spray bottle.

Just like mopping floors, mixing an insect growth regulator (IGR) with Nibor-D will make it more effective. Mix NyGuard IGR at a rate of 0.4 ounces (12 ml) to one gallon (3.8 liters) of mixed Nibor-D and water.

For small jobs, you can mix Nibor-D at a rate of 1/3 scoop to 16 ounces (473 ml) of cold water.

Also, for small jobs, you can mix 1.5 ml of NyGuard IGR with 16 ounces (473 ml) of mixed Nibor-D and water.

4. After treating the floors, you need to treat other surfaces in your home. Mix Nibor-D at a rate of one scoop of product to 16 ounces (473 ml) of cold water in a small bucket and apply it with a sponge or pour it into a handheld spray bottle. The spray bottle is way faster, but a sponge is a lot neater. You may want to do both. Spray on the liquid and then wipe off the excess with a sponge that is wet with the mixed Nibor- D.

Like before, you can also mix NyGuard IGR with the Nibor-D. The correct dilution rate is 1.5 ml of NyGuard IGR with 16 ounces (473 ml) of mixed Nibor-D and water.

Apply this to all surfaces inside all your cabinets. This would include the inside of all shelves and all surfaces that cannot be seen from the outside of the cabinets. **The only exception is that you should not apply Nibor-D on surfaces where food, dishes, or utensils will come in contact with it.**

You need to remove the drawers from all the base cabinets and apply this on the outside of all the drawers and all surfaces inside the base cabinet. Don't forget to treat the bottom side of the sinks and all the pipes.

Also apply Nibor-D under all small appliances. Just don't apply it on surfaces where food is prepared, served, or stored.

5. While the drawers are removed, also apply some boric acid or diatomaceous earth (DE) on the floor under the base cabinets. If the cabinet has a bottom, drill a hole in it and apply boric acid or DE. If you can remove the dishwasher, then you should apply Nibor-D on all surfaces on and around the dishwasher, and you should apply boric acid or DE on the floor under the dishwasher as well.

The boric acid can be applied with a handheld duster. Personally, I think the easiest duster to use, and the one that does the best job is the one that is shown in Figure 3. This duster is available from Amazon, and it is quite inexpensive.

Figure 3 - Handheld duster by TRIWISH

6. Next, you need to apply D-Fense dust in the pipe penetrations that go through the wall and the bottom of the cabinets. I recommend buying a separate duster for each formulation of dust that you use.

7. Remove all outlet and switch covers and apply a little D-Fense dust on the bottom of the outlet boxes. You should also insert the duster tube between the outlet box and the drywall and puff some dust into the wall void. You can apply this dust with a handheld duster as well.

8. After treating the kitchen and bathroom cabinets, you need to treat the sides of all appliances with Nibor-D. You do not need to treat the areas of the appliances that you can see. You also need to treat all surfaces surrounding the appliances and the floors with Nibor-D. Just like with the base cabinets, apply boric acid or diatomaceous earth on the floors under the appliances, after the Nibor-D has dried.

9. Apply Temprid FX on all kitchen and bathroom baseboards. See Figure 4. Mix this product 8 ml to ½ gallon (1.9 liters) of water. This product can be easily applied with a handheld spray bottle or pump-up garden sprayer. Pump the garden sprayer a couple of times to produce a light fan spray.

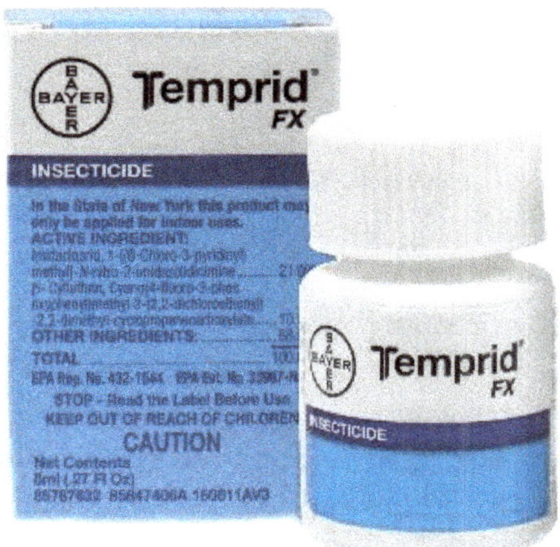

Figure 4 - Temprid FX

10. Select a good, professional bait and apply it in and under all cabinets. Also apply the bait around all appliances.

When applying bait, only use pea-size dots, and do not put them any closer together than 12 inches (30 cm).

The bait needs to be changed every 90 days to a bait that contains a different active ingredient because roaches will become resistant to each active ingredient after that period of time.

A list of baits can be found at the end of this book and at http://www.guyspestsolutions.com.

11. Roaches will often hang out in drains. The easiest way to treat drains is to shoot a little Shockwave 1 down the drain. See Figure 5. This will not harm pipes or septic tanks. In my opinion … while the little bit of pesticide you shoot down the drain will kill all the roaches; it will not harm the environment at all. So, I would not fret about that.

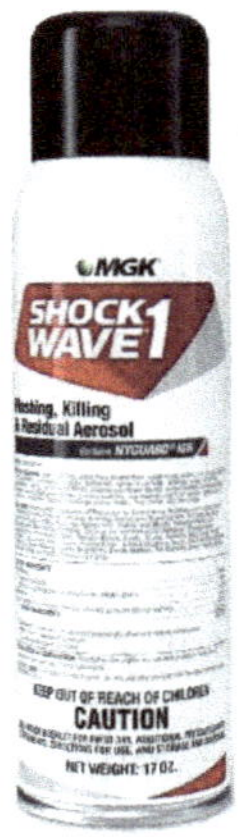

Figure 5 – Shockwave 1

By the way, we did not enter your home through the drains. It is impossible for us to get past the drain trap under the sink. We are in your drains because there is water and drain sludge there. We like water and drain sludge, so we entered the drains from inside your living space to obtain these materials. Drains also offer a dark hiding place for us.

12. Finally, you need to put a Gentrol Point Source IGR disc under each of the appliances, one under the sink, and one in the cabinets, spaced at five-foot (1.5 meter) intervals. See Figure 6.

The disks MUST be replaced every 90 days for an entire year without fail. Mark the date on the disks with a Sharpie and place a note on the refrigerator with the date when the disks need to be replaced. **This is very important.**

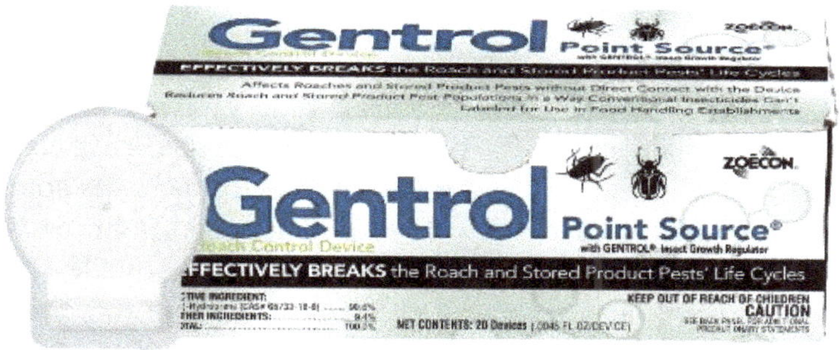

Figure 6 - Gentrol Point Source Disks

13. Most pest control supplies can be obtained from Amazon, domyown.com, solutionsstores.com, and other pest supply stores. There are also links to all the products on my website at www.guyspestsolutions.com. Please note that some pest control supplies cannot be shipped to some States and countries that restrict the sale of pesticides to licensed operators. I have devoted an entire chapter explaining your options if you cannot obtain the recommended products. While you will always achieve the best results by using all the recommended products, it is possible to greatly reduce or eliminate an infestation by modifying the procedure. You should also know that, if one supplier will not ship to you, another supplier may not have a problem with it.

Note: If you cannot obtain Temprid FX, just skip using it and apply the mixture of Nibor-D and NyGuard IGR to the baseboards in the kitchen and bathrooms instead. Unfortunately, imidacloprid, the active ingredient in Temprid FX, has been restricted to licensed applicators (pest controllers) in some States and many countries around the world. While well meaning, these laws were not very well thought out. Nevertheless, those laws do exist, so we need to figure out a way to deal with the problems that those laws have created. Don't worry ... Nibor-D will still get the job done, albeit somewhat more slowly.

9 PHASE 2 – FINISH US OFF

Now that you have taken steps to quickly reduce your cockroach population, it's time to finish us off. To do that you are going to need to bring the battle to us. Like I said before, about 90% of us never come out of hiding. If you plan to wipe out the rest of us by relying only on stupid roaches to bring poisons back to us, then you are destined to fail. That will never work because those of us who survive will now be resistant to all the pesticides you used. We will be stronger than ever. After that, it won't take long before we multiply in numbers that would frighten even Stephen King. If you have been using a professional pest controller, or you have previously applied the Phase 1 products, then you may have already created an army of super roaches. In that event, Phase 1 may not achieve the desired results as quickly as you would like. Fear not … all is not lost. Phase 1 will still work, albeit somewhat more slowly and Phase 2 will still finish us off.

The only way to ensure that we all die is to find a way to drop a pesticide directly into our hiding places. Since we do not come out of hiding anyway, the best product to do this is a permanent pesticide, such as boric acid or diatomaceous earth (DE).

Let me be perfectly clear. We ARE breeding in the walls. The way we got in there was we entered through electrical outlets, switches, and penetrations that were made in the wall to run the pipes from the sinks into the walls. Now that we are in the walls, we can travel at will from wall void to wall void through holes that were created in the studs by the electricians. That means we can exist in virtually any wall void in the entire structure. The problem is that you will never know which wall voids we are occupying. Obviously, we do not occupy all of them, but if you do not treat all of them, then you may miss some of the voids that we are using as a breeding ground. For this reason, you must treat all the wall voids.

To do this, you need to drill a hole at the top of each wall void and blow in boric acid, or DE. To find the wall voids, you need to use a stud finder. You can get a variety of different stud finders, but the one I like best is the Franklin Sensors ProSensor 710+ Stud Finder. See Figure 7. This is available from Amazon and some hardware stores.

Figure 7 – Franklin Sensors ProSensor 710+ Stud Finder

Step-by-step procedure:

1. Now that you have a stud finder, get a ladder and locate all the studs along all the walls, and make a pencil mark at the center between each stud. You do not want to drill into the studs, but rather the hollow space that is next to the studs. If you drill into wood, then you are drilling in the wrong place. Please note that, most of the time, studs are spaced 16 inches (41 cm) apart, so you should have holes located about every 16 inches.

2. Make a pencil mark at the center of the wall void that is about 12 inches (30 cm) below the ceiling. See Figure 8. Drill a hole, at a steep, downward angle, that is wide enough to allow the tip of the duster to enter. The reason you drilled the hole at a downward angle was to avoid any insulation that may be in the wall.

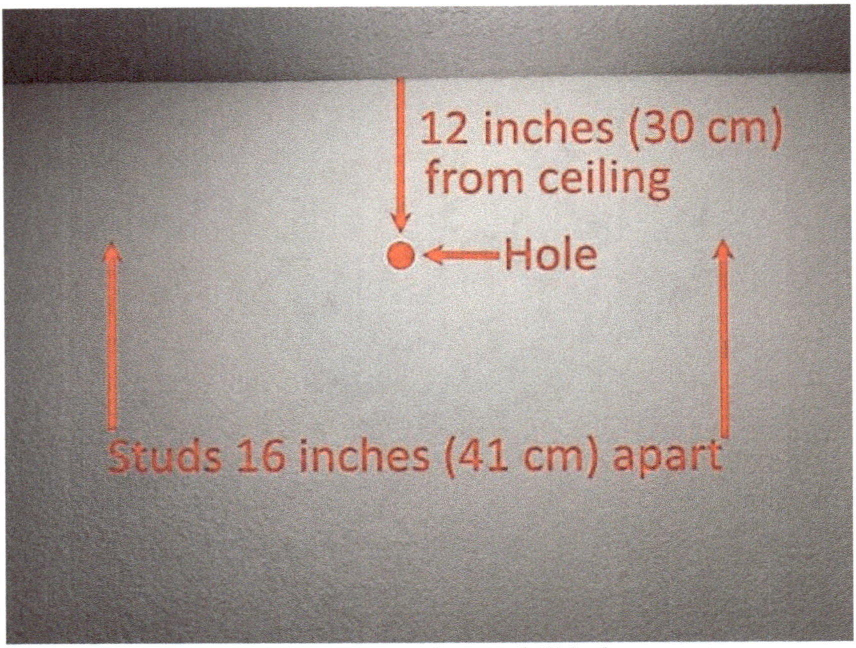

Figure 8 – Location to drill holes

3. You need to blow at least one tablespoon of boric acid or DE into each void. The best way to do this is with an electric duster. Professional electric dusters are expensive, so I recommend that you make one yourself out of a cheap air pump that is made for inflating air mattresses and pool toys, and an empty peanut butter jar. Full directions on how to make one can be found in Chapter 17.

4. When you are treating wall voids, keep in mind that you also need to treat above and below windows and above doors.

Also, keep in mind that the studs are not always the standard 16 inches (41 cm) apart. Studs can sometimes be an inch (2.5 cm) or less apart, so you need to be careful that you find all the voids, even if they are only half an inch wide. If you fail to treat a void, the roaches may very well find it.

5. Don't forget to treat the wall voids behind the kitchen cabinets. To do this, remove all the dishes and other items from the kitchen cabinets. Look for screws that secure the cabinets to the wall. The location of that screw is most likely also the location of a stud that is behind the cabinet. This is not always the case, but oftentimes it is. Drill a hole next to that screw that is wide enough for the duster tube to enter, but do not drill it all the way through the wall. There may be a gap between the back of the cabinet and the wall where cockroaches may be breeding. If that is the case, you want to drill the hole through the back of the cabinet, without drilling it through the drywall. Once you have done this, apply the dust into this void. You may need to use a hand duster to treat that void. Next, drill the hole through the drywall and apply dust into the wall void using the electric duster. If you hit a stud, then drill a new hole a couple of inches to one side of the original hole. After you finish, just put a small piece of duct tape over the hole. Pro tip: To make the hole less noticeable, remove a shelf and drill the hole directly behind where the shelf was. When you replace the shelf, it will hide the duct tape.

Holes in the drywall can be repaired with spackling compound. This can take a little practice, so start in a closet and practice on the same hole until you have a job that will make you proud. There are YouTube videos that can show you how to do this like a pro.

6. If you live in a two-story home, then you also need to treat the ceiling above the first floor. You do this much the same way as treating the walls. Use the stud finder to locate the ceiling joists. Just like with the walls, drill a hole between the floor joists that is large enough to insert the duster tube. This hole does not need to be next to the walls. In fact, it would be best to drill it somewhere in the center between both walls.

Sometimes there are cross members (bridging) between the floor joists that will stop the dust from getting across the entire void. So, you always need to run the stud finder along the center of the void, from one wall to the other, to check for bridging. Newer homes typically use cross bridging or X bridging which are more like brackets. Since these are not solid, they will allow the dust to pass by. It is typically older homes where you will find solid bridging. Obviously, if you have solid bridging along the void run, then you must treat the void on both sides of the bridging. The best way to do this is to drill your hole in the center between the wall and the bridging. If there is no bridging, then you may get away by just drilling a hole at the center point between the two walls. If it is a very large room, then it may be wise to section off the length of the void into thirds and drill two holes. The idea is that you want to ensure that you have a layer of dust along the entire length of the void.

7. German cockroaches usually do not hang out in the attic, but that does not mean we can't. Therefore, it is a good idea to treat the attic as well. To do this, use a small shop vac that has the ability to attach the hose to the exhaust port. Remove the filter, attach the hose to the exhaust port, and put several cups of boric acid or diatomaceous earth (DE) in the vacuum. Simply turn on the vacuum and blow the dust across the entire attic. Always make sure that you wear a respirator, eye protection, rubber gloves, a long-sleeved shirt, and long pants.

If you have a situation where the attic floor is covered with plywood or some other material that would make it impossible to treat the ceiling void by blowing dust over the top of the insulation, then there really isn't much you can do to treat those voids. You could drill a lot of holes through the flooring in the attic and blow dust on top of the insulation, but since it is somewhat unlikely that this area will be infested, it's probably best not to treat it at all.

Instead, treat the ceiling below the attic by wiping Nibor-D, mixed with NyGuard IGR, onto and around all ceiling fixtures, such as lights, fans, and so forth. After it dries, puff boric acid or DE in all cracks and crevices in the ceiling. The idea is that you want to force the roaches to cross this barrier when they try to leave the ceiling void. The NyGuard IGR may stop them from reproducing, thereby ending the infestation in the ceiling.

8. If you have a crawl space, it is unlikely that you are going to find German roaches that have taken up residence there. However, you cannot rule that out entirely. It does make sense to go into the crawl space and inspect for roach activity. If you find that German cockroaches are living in the crawl space, then you must treat all surfaces in the crawl space with Nibor-D. This would include the walls, the piers, the floor joists, and the subflooring. Treating the joists and the subflooring will require that you remove the insulation prior to the treatment. After you have finished treating, then reinstall the insulation. If you have foam insulation, just treat over the top of it. Mix the Nibor-D at a rate of 7.5 scoops per gallon (3.8 liters) of cold water. For best results, add 12 ml of NyGuard IGR to one gallon of mixed Nibor-D.

9. Now that you have treated all the main breeding areas, you need to treat all the baseboards with Nibor-D, the same way that you treated the inside of the cabinets. That is, mix the Nibor-D at a rate of one scoop to 16 ounces (473 ml) of cold water. A 4 fluid ounce measuring scoop comes with the product. You can apply this product with a sponge or a pump-up garden sprayer. If you want to mix a gallon of product, then mix it at a rate of 7.5 scoops per gallon (3.8 liters) of cold water. Remember … do not mix more than you are going to use within two hours.

For best results, also add NyGuard IGR at a rate of 1.5 ml to 16 ounces (473 ml) of mixed Nibor-D, or 12 ml of NyGuard IGR to one gallon of mixed Nibor-D.

10. After treating all the baseboards in the entire living space with Nibor-D, then let it dry and apply boric acid or diatomaceous earth under all the baseboards. If the baseboards are sealed to the floor, then you do not need to do this step.

11. Now apply Nibor-D to all surfaces that are under electronic equipment. You know … things like TVs, computers, gaming systems, and so forth. You cannot treat inside electronic equipment, but by treating under these items, you will prevent roaches from coming and going without dying. Use the same dilution rate that you used for treating cabinets and baseboards. Adding NyGuard IGR is always a good idea.

12. Put a pea-sized dot of bait on a piece of paper and place one next to each piece of electronic equipment.

13. Place a Gentrol IGR disk next to each piece of electronic equipment. Make sure that you replace them every 90 days. Failure to do this may result in your failure to eliminate the infestation. Of course, I would be perfectly okay with that outcome. I mean, I am agreeing to share these secrets, but that doesn't mean I want you to be successful.

14. In the bathroom, remove the toilet tank lid and apply the same mixture of Nibor-D and NyGuard IGR that you used for other surfaces inside the tank, from the top of the tank to the waterline. Apply boric acid or diatomaceous earth (DE) in the gap between the toilet tank and the bowl using a handheld duster.

15. Next, go around the entire interior of the living space and look for any cracks or crevices that you can find, and apply boric acid dust or DE into all those cracks and crevices, using a handheld duster.

16. Remove all the outlet and switch covers and apply boric acid or DE on the bottom of the electrical boxes. Then insert the duster in between the outlet box and the drywall and apply some dust into the wall void. I know, you already blew boric acid or DE into all the wall voids, but it doesn't hurt to do it one more time.

When applying dust into the outlet boxes, you do not want to apply it directly into the outlet fixtures themselves. If there are roaches in the outlets where the plug goes in, then buy some child safety plug covers. These will prevent the roaches from getting in.

You may recall that in Phase 1 you applied D-Fense dust in the kitchen and bathroom outlet and switch boxes. Since this product only has a nine-month residual, you must retreat the same boxes with boric acid or DE.

17. Treat the ceilings around all ceiling fans and light fixtures with Nibor-D. You can use a sponge to lightly apply it directly on ceiling fans and light fixtures.

18. Now inspect all the doors in the entire living space for roach activity. This would include things like fecal matter, dead bodies, live roaches, and so forth. If you see any evidence at all of cockroach activity, then apply Nibor-D mixed with NyGuard IGR on all surfaces of the doors.

19. You need to treat all the furniture in your living space. Apply Nibor-D mixed with NyGuard IGR under the kitchen and dining room tables, where it cannot be seen. Also, treat the bottom side of kitchen/dining room chairs. Do not treat areas that may come in contact with people or pets.

20. Treat bedroom dressers inside and out. Remove the drawers and treat all surfaces inside the dresser. Also, treat all the exterior surfaces of the drawers. Be sure to treat the back side of the dresser and make sure that you have the floor treated under the dresser as well. As always, if it looks unsightly, don't treat the visible surfaces. Treat nightstands and other such furniture in the same manner. Do not treat surfaces where clothing will be stored.

21. Remove everything from bookshelves and treat all surfaces of the shelves with Nibor-D. Let it dry and then apply boric acid or DE to all cracks and crevices. Then replace the books or other items.

22. For upholstered furniture, spray Nibor-D on the bottom side of the furniture, and place a Gentrol IGR disk under each piece of furniture. Make sure that the floor under the furniture is treated with Nibor-D. As always, don't forget to change the Gentrol IGR disks every 90 days.

23. Spray Nibor-D on the back side of all wall hangings.

24. Apply Nibor-D to the back side of all draperies.

25. The bottom line is that, without exception, you want to apply Nibor-D on all surfaces that are not visible. Also apply it on visible surfaces, if you can tolerate looking at it. Apply boric acid or DE everywhere it will not be disturbed by people, pets, or air currents. Be thorough and leave no stone unturned.

Please note that, if you are still seeing cockroach activity 6 months after treatment, then you missed something. There may be a wall void, a space behind a door jamb, and that sort of thing, that you failed to treat.

Also, keep in mind that it takes a full year for the IGR disks to do their job, so you may occasionally see a roach near an appliance or piece of electronic equipment. Don't worry … these guys will die before they can get into any mischief.

Did I mention that you MUST change the IGR disks every 90 days if you want to successfully eliminate the infestation? I can't say it often enough. It's that important.

10 APARTMENT MODIFICATION

Of course, the best way to treat a German cockroach infestation in an apartment is not to have one in the first place. Landlords will often fog apartments to fool unsuspecting tenants into thinking there is no cockroach infestation. Do not be fooled by this trick. There are ways to determine if the apartment you wish to rent has an existing infestation of German cockroaches.

The easiest way to do this is to ask the neighbors. I know that most of you humans hate to interact with each other, especially folks you don't know, but that will be a cause for your undoing. We roaches have no such reservations about interacting with each other. So, take a lesson from us. After you have finished looking at the apartment, return that evening and knock on your neighbors' doors. Tell them that you are planning to rent an adjoining apartment, and you are wondering if they are having any problems with roaches. This is also a good time to ask them if they have any other issues with the landlord. You would be amazed at how willing humans are to share this information.

If for some reason speaking with the neighbors is not an option, then there are ways you can check the apartment for roaches. The cleaning crew will inevitably miss certain areas. You need to look for fecal matter, dead bodies, body parts, and of course, live roaches. Start with the obvious places. Use a good flashlight to check the inside of the cabinet hinges. Look at the outlets and switches in the kitchen and see if you can find any fecal matter. It will look a bit like black pepper stuck to the outlet and switch covers or on the wall. See Figure 9. If you see anything like that, then run. This is not the apartment for you.

If those areas look clean, then look under the kitchen sink. See if there is cockroach fecal matter on the pipes or the bottom of the sink. Don't be afraid to stick your head right under the sink and look up. Cleaning crews rarely clean the bottom of the sink.

If the kitchen sink comes up clean, then inspect inside the closets. You need to look at the inside of the closet between the door jambs and the walls. Housekeepers rarely get into that area that is right next to the door, inside the closet.

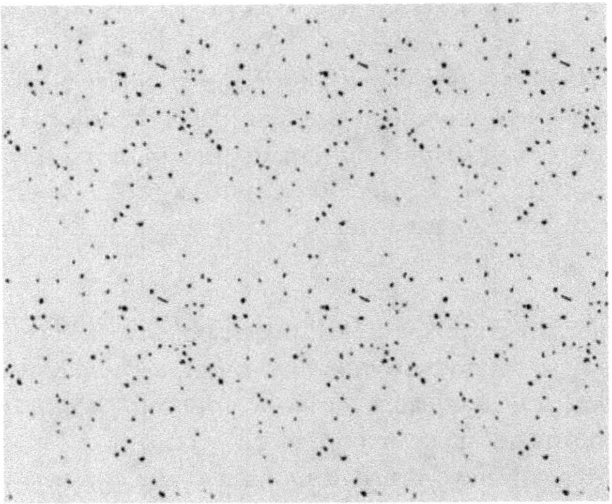

Figure 9 - Cockroach fecal matter

If the closets come up clean, then pull out the refrigerator. Oftentimes housekeeping will forget to clean under and behind the refrigerator. If they do not allow you to pull out the refrigerator, then red flags need to go up in your mind. Why wouldn't they want you to look behind the refrigerator? If this is the case, use your flashlight to look between the sides of the refrigerator and the cabinets or walls. Also, open the doors and look at the door seals. Many times, the cleaning crew will forget to clean the refrigerator door seals.

If the refrigerator comes up clean, then pull out the stove. If you cannot pull out the stove, then use the flashlight to have a good look behind and under the stove. Please note that some stoves are difficult to pull out, so there could be a good reason why they do not want you to do this. Refrigerators are usually on wheels, but stoves are not. Also, stoves may be hooked up to gas lines and the person showing you the apartment may be somewhat nervous about moving the stove. You can look inside the oven, and some stoves have a lid surrounding the burners that can be easily lifted. Cleaning staff often forget to clean under the burners.

Again, you are looking for fecal matter, dead bodies, body parts, and of course, live roaches. If you don't find anything like this, then the apartment is probably okay to rent.

Even though you did a thorough inspection of the apartment, you may miss something, or you may get an infestation after you move in. A German cockroach infestation is going to be more problematic for an apartment than for a single-family home. If you have German cockroaches, then the apartments surrounding you also have them, and I can assure you that we are breeding in the walls as well.

Since the landlord is probably not going to allow you to drill holes in the walls, and since we will continue to come in from adjoining apartments, it is virtually impossible to eliminate the problem 100% in an apartment. Still, all is not lost. If you follow the rest of the treatment plan, except drilling holes in the walls, you may never see another cockroach again. If you do a thorough job with Phase 1 and 2, and you apply Nibor-D and boric acid (or diatomaceous earth) in every place a cockroach can possibly place its foot, then we will have no choice but to cross it when we enter your apartment. Contact with these materials will result in death within several days, and that is not enough time for us to reproduce. Therefore, we will no longer be able to establish an infestation in your apartment. When we enter the apartment, we will generally seek a hiding place as quickly as we can, so it is possible that you may never see us. If you do see one of us, you can be confident that we will be dead within a few days. Remember… the key to achieving this is thoroughness. Leave no stone unturned.

Finally, after you are finished with Phase 2 in the kitchen and bathrooms, be sure to seal up all cracks and crevices around pipe penetrations. There is no sense in making it easy for us to get in from another apartment. A great way to do this is with plumbers' putty. You can get it at Lowes or Home Depot in the plumbing department. It kind of works like Play-Doh. You can mold it into any shape you like and squeeze it into cracks and crevices. It's also great for sealing up outlet covers and switch plates. Just remove the cover and put a bead of putty around the outlet cover to seal any places where roaches can get through. When you put it back in place, the excess will squeeze out, and you can easily wipe it off and reuse the excess. It never hardens, so the covers can be easily removed for maintenance. It's pretty cool stuff and it is very inexpensive too.

Condominiums, townhouses, and row houses have the same problems as apartments and can be treated the same way. If you own the property, then drill the walls. Otherwise, just treat it the same way as an apartment.

11 MODIFICATIONS FOR THOSE WHO CANNOT BUY PESTICIDES

Many countries around the world and some States in the US have restricted the sale of pesticides to licensed pest controllers. Even though these laws are unwarranted and make no sense, they do exist, so you must deal with them. . Obviously, treating a German cockroach infestation without the proper products is going to be much more of a challenge, but I will do the best I can.

Most of the time, it is still legal to buy boric acid and food-grade diatomaceous earth (DE) in just about every country or State. You can mix DE with cold water and apply it everywhere I told you to use Nibor-D. I know I told you earlier that mixing DE with water was not going to work as well as Nibor-D. But that was only because DE has not been properly field tested for use on roaches when it is mixed with water. So, the effectiveness of this has not been proven. Also, when a DE and water mixture dries, the DE that is left behind will rub off the surface much more easily than Nibor-D. This makes treating floors more problematic because it will come off on your feet. Therefore, you may only be able to treat areas under furniture and next to the walls. If this is the case, then treat under upholstered furniture and leave a 6-inch (15 cm) band around the furniture. Also, create a 6-inch (15 cm) band around the walls.

Despite these issues, DE is the best shot you have at this point, and I can tell you that you won't catch me dead walking on it. Did you see that little play on words there? Just a little cockroach humor. This procedure will leave a powder behind and that is what will kill us. How much you dilute it will depend upon how much you can tolerate with regard to how it looks. It will work best if you mix 1 part DE to 4 parts of cold water by volume. However, this is going to look quite unsightly. So, you can try mixing 1 part DE to 6 parts water and see how that looks. If it's still too nasty, then try mixing it 1 to 8. It should still work, albeit not quite as effectively. The idea is that you need it to leave behind a light dust. The heavier the dust, the better it works. If it looks too unsightly, just avoid applying it on surfaces that you can see, like the outside of cabinets. This is going to leave a film on everything you apply it on, but you should know that, when you are ready to move out, the DE can be easily washed off.

The best way to apply DE is by wiping it on with a sponge or rag, but you can also pour it into a handheld spray bottle. Using a sponge or rag will give you a more even coat that will not look quite as bad. If you like, you can spray it on and just smooth it out with a sponge or rag that is wet with the DE mixture.

Be sure to apply the DE on every possible surface imaginable that will not be damaged by water, unless it really looks bad. The key to eliminating German cockroaches is thoroughness. You must apply a pesticide on every single place where a cockroach can possibly place its feet. I know that it is horrible to live with DE covering just about everything, but it sure does beat living with us. At a minimum, apply it inside cabinets, on baseboards, and any other place a roach would need to walk to move to another area. The more places you treat, the less likely the roaches will be able to escape the treatment.

After the DE dries, you need to apply boric acid or DE **dust** in all the places previously mentioned in Chapter 9. Basically, you want to apply boric acid or DE dust to virtually every place where it will not be disturbed by people, pets, or air currents. Don't forget that you need to apply DE in all the wall voids.

Please note that you should not mix boric acid with water, and you cannot apply it in places that are wet because that may render it ineffective. Also, apply the boric acid or DE dust AFTER you do all the spraying with liquids. If you live in an apartment where you cannot drill holes in the walls, just skip that part of the treatment.

After you are finished applying boric acid and DE, then you should also apply a bait. Obtaining a good bait may be difficult where you live. If you cannot obtain any commercial baits, you can try making your own bait. Homemade baits are nowhere close to being as effective as commercial baits, but they are better than nothing. Try mixing equal amounts of sugar and DE with a little water to create a paste. You can do the same thing with boric acid. I know I said that you cannot mix it with water, but creating a bait is an exception because the roaches will be eating it instead of walking through it. I do not think this paste will stick well to surfaces, so you may need to simply put a little bit of it on pieces of paper and place them strategically where you see cockroach activity.

You may need to experiment with the formulation to make it the most appealing to the roaches. That is to say, if they are not eating it, try adding more sugar. Some people add an egg yolk to the mixture, which may help, but I think that may also create some unwanted odor. I know of some humans who are experimenting with mixing boric acid with peanut butter. I don't know if that will work, but it has the advantage that you can apply it to vertical surfaces.

Never apply any type of pesticide over the top of bait because that will prevent the cockroaches from eating it. That is why you always use the bait last. The correct order of application is always liquids first, dust second, and bait last.

The biggest problem you will have is treating appliances and electronic devices. You need the Gentrol IGR disks to treat these items because you cannot treat the inside of them with DE. Unfortunately, I do not have a good solution for that. The best you can do is to apply the DE under and around these items. This will prevent the roaches from traveling to other areas without dying.

12 AN OUNCE OF PREVENTION

I'm sure you recall that old adage that "An ounce of prevention is worth a pound of cure." Well, that definitely applies to German cockroaches. It is much easier to prevent an infestation than it is to eliminate one. Believe it or not, there is a way to permanently roach-proof your home or business so that you can never become infested with German cockroaches. To do this, please refer to Chapter 8 and 9, and familiarize yourself with Phase 1 and 2. Essentially, what you need to do is the entire treatment, with a few exceptions. You do not need to drill the walls because you do not have any roaches in the wall voids. Also, you do not need to use bait or an IGR, such as NyGuard IGR or the Gentrol IGR disks.

However, you must treat areas where cockroaches are likely to enter the environment. Remember that we do not live outside, so we must be carried in by people who encounter us in another building. Also, keep in mind that we cannot fly. Therefore, we are most likely going to enter your living space, either from a horizontal surface, such as a countertop or seating, or from the floor.

You may recall that we are hitchhikers. Most of the time, we will either be in someone's belongings, or on their clothes. Occasionally, we may also enter your environment because you brought in a package from a store, or a package was delivered in which we were hiding. This includes appliances and electronic equipment, such as TVs, computers, gaming systems, and so forth. Any person or item that enters your home has the potential to deliver a female roach or an egg sac. While you are less likely to become infested by new appliances and electronics, it can still happen. So, you should always be on your guard for possible cockroach hitchhikers.

If you understand this, then you can see that a German cockroach infestation can be prevented by applying a permanent pesticide on all surfaces that we would need to cross to reach a hiding area.

If you are wondering under what circumstances you should roach-proof your home, the answer is simple ... everyone should do it. Since it is very easy to become infested with German cockroaches, taking this step is always a good idea. If you live in a multi-unit building like an apartment, condo, townhome, row house, etc., then this is a must if you do not want to become infested. Sooner or later, someone in the building is going to introduce German roaches into their unit. After that, it is simply a matter of time until the infestation spreads throughout the entire structure. Any time you have adjoining walls or floors you are at a much greater risk for an infestation than if you live in a single-family home.

If you own a restaurant, then it is also very important to do a preventive treatment. This is especially true if your business shares walls with other businesses or apartments.

Commercial real estate, such as hospitals and hotels, is also at high risk for an infestation.

Step-by-step procedure:

1. Start by treating all baseboards with Nibor-D and applying boric acid or diatomaceous earth (DE) under all those baseboards after the Nibor-D dries. You may recall that you should mix 1 scoop of Nibor-D with 16 ounces (473 ml) of cold water for this application. You can wipe it on with a sponge or apply it with a spray bottle. If you are treating a lot of baseboards, you can mix Nibor-D at 7.5 scoops per gallon (3.8 liters) of cold water and apply it with a pump-up garden sprayer. It is not necessary to add an insect growth regulator for a preventive treatment.

2. Now, treat the inside of all the cabinets in the kitchen and bathrooms with Nibor-D, mixed at the same dilution rate. After it dries, apply boric acid dust or DE in all pipe penetrations and in any cracks and crevices that are in or around cabinets.

3. Next, apply Nibor-D on the outside of appliances, where it cannot be seen. There's no need to treat the front of the appliances or any areas that are visible. However, you must treat the back and sides of appliances and surfaces that surround them. This would include the floors, sides of cabinets, walls, and so forth. When that dries, apply boric acid dust or DE under all the appliances.

4. Apply Nibor-D to the inside of all cabinets in the kitchen and bathrooms and on the outside surfaces of all drawers, as I explained in Chapter 8.

5. Even though you do not have roaches in the wall voids, you should still apply boric acid or diatomaceous earth (DE) in the pipe penetrations under the sinks in the kitchen and bathrooms. It is a good idea to remove all the outlet covers and apply boric acid or DE on the bottom of the outlet boxes and between the electrical box and the drywall. You should do that with the wall switches as well.

6. Make sure that you apply Nibor-D on the bottom side of all furniture and on the floor under all furniture. The only exception to applying Nibor-D on the floor under furniture is that you do not need to do it for items where the floor can be easily seen. For example, you do not need to treat the floor under the dining room table and chairs.

7. It's a good idea, if you live in an apartment, to also apply a band of Nibor-D on the door jambs of the entrance door and on the floor between the door jambs. It is not uncommon for us to simply enter your apartment from the hall. Placing this barrier will force cockroaches to cross it on their way in, and they will die within a few days. Make sure that you treat the entire door jamb, including both sides and the top.

8. As always, if you remove the Nibor-D, boric acid, or DE, then you must reapply them to maintain protection.

13 MOVING

If you are moving out of an apartment or home that is infested with German cockroaches, then you certainly do not want to bring any with you, because that will start an infestation in your new home. The problem is, this is way easier said than done. The reason German cockroaches spread so quickly is because we will get into anything we can, and we will even latch onto your clothing. Therefore, it is a safe bet that some of us are in your appliances, electronics, furniture, clothing, and maybe even your books. So, let me go over these items one by one and explain what you need to do. Keep in mind that, no matter how meticulous you are, there is always a possibility that you missed an egg sac or female roach. The sad truth is that the only sure way to prevent us from coming along with you is to throw away all your belongings. Since most people are reluctant to do that, I will explain the next best thing. If what I am about to tell you seems excessively radical to you, that is because it is, but that is exactly what it is going to take to significantly reduce the possibility of infesting your new home.

Step-by-step procedure:

1. Needless to say, the first thing you want to do is to throw away everything you can. If it is not an item of important sentimental value and it is something you can afford to replace, then I would recommend getting rid of it. This is particularly true when it comes to upholstered furniture. It is very difficult to treat all the areas of an upholstered couch or recliner where cockroaches may be hiding.

2. Upholstered furniture: If you have no choice, and you must keep your couch or recliner, then treat this furniture in isolation.

Rent a storage facility that is not climatically controlled. You want it to get hot and cold. Purchase a king-size plastic mattress bag that is used to store mattresses and see if you can get the couch or recliner inside of it so that it can be sealed up. You can often find these at places like U-Haul. Sometimes you can even find a bag made just for a couch or recliner. Before placing the furniture in the bag, you need to treat it, but don't treat it until you get to the storage facility. You are going to treat the storage facility much the same way as you would a house or apartment.

Before bringing the furniture into the storage facility, make sure that you have treated the entire floor and walls of the storage facility with Temprid FX. You should mix this product at a rate of 8 ml to ½ gallon (1.9 liters) of water. You should do the same thing with the truck you are using to move the furniture. Before putting the furniture into a truck, the entire bed of the truck and the walls should be treated with Temprid FX. This way, the truck will not become infested. It's the same with the storage facility. If you treat the floor and walls with Temprid FX, then any cockroaches that escape from the furniture are going to die before they can reproduce in the storage facility. Let's face it, it is not very nice to allow the storage facility or the truck to become infested with cockroaches. You wouldn't want somebody doing that to you, so do not do it to them either. If you cannot purchase Temprid FX, then do the same treatment with Nibor-D, mixed with NyGuard IGR. Use the same dilution rate that you used for cabinets.

Now that you have the storage facility treated, remove the cushions and treat them first. If there are zippers on the cushions, then remove the covers, spray Nibor-D on the cushions, and then replace the covers. Mix the Nibor-D at a rate of 1/3 scoop to 16 ounces (473 milliliters) of cold water.

Purchase at least one can of Nuvan Directed Spray Aerosol and shoot it into every crack and crevice in the entire couch or recliner. See Figure 10. **VERY IMPORTANT! Always wear a respirator that is approved for vapors, rubber gloves, and eye protection when working with Nuvan Directed Spray Aerosol.** Shoot it into all the seams where the cushions were, then flip the couch or recliner over and inject it into every space possible, even if you need to make small holes in coverings under the furniture. A bit of caution is required here because you want to shoot in enough to get to the cockroaches, but not so much that you damage the furniture. You don't want to get this material on any visible areas, so don't use so much that it soaks through the fabric. This can be a little tricky, so use your best judgment to prevent damage. You may want to test this product on an inconspicuous area before use. After you have done this, then apply a bait to the underside of the furniture, in several areas.

Figure 10 - Nuvan Directed Spray Aerosol

Immediately place the furniture inside the plastic bag and toss in a Gentrol IGR disk. Make sure that the bag is completely sealed off, so that it is airtight. If there is the slightest opening, then seal it off with duct tape.

Leave the furniture in the storage facility for at least 100 days, which is the average life cycle of a German cockroach. After that time, you can go in and look through the bag to see if you can observe any evidence of cockroach activity. Look for dead bodies, body parts, fecal matter, and of course, live cockroaches. If you see anything like this, then you need to remove the furniture from the bag, re-treat it, and then re-bag it with a fresh IGR disk. Try opening it again in another 100 days. If you do not see any activity, then remove the furniture from the bag and inspect it very carefully. If everything looks good, then go ahead and bring it home. Otherwise, you need to repeat the process for another 100 days. The roaches should not be able to survive this for more than a year.

3. Large appliances: Things like refrigerators and stoves are expensive, but they are also very difficult to treat. Therefore, bringing large appliances with you will almost certainly guarantee that you will be infested in your new home.

There is no easy and effective way to treat large appliances. All it takes is one egg sac and you will have another infestation. If you must take the appliances with you, then you need to assume that you are bringing roaches with you as well. So be sure to set up a barrier in the entire home before you bring in the appliance. This way, when they attempt to leave the appliance, they will cross a kill zone and die. Apply all the treatments outlined in Chapter 12 and then perform the following treatments as well. Some of this is already covered in Chapter 12, but it's worth repeating.

Mop all floors with Nibor-D, and create a 6-inch (15 cm) band on the carpets around all walls with Nibor-D before moving in.

Spray or wipe Nibor-D on any surface of the appliance that will not be damaged by water.

Apply a bait in several areas on the back and sides of the appliance where it cannot be seen.

Apply Nibor-D on all wall and cabinet surfaces where the appliance is being installed. Also, apply Nibor-D on the floor under the appliance.

Apply boric acid or diatomaceous earth (DE) under the appliance.

Put a Gentrol IGR disk under the appliance and replace it every 90 days for a full year.

4. Small appliances: Again, this certainly bears repeating. If you have a small appliance that you can get rid of, that would be the way to go. However, if you cannot get rid of that coffee maker or blender, then follow these procedures.

To the greatest extent possible, disassemble the appliance and clean it just as well as you can.

Spray or wipe Nibor-D on any surface that will not be damaged by water.

Apply a bait on the bottom of the appliance.

Place the appliance in a sealed trash bag, along with a Gentrol IGR disk.

Do not open the bag for at least 100 days. When it is time to open the bag, you should always do it outside. If possible, open the bag on top of concrete, like a driveway or sidewalk. Before opening the bag, make sure you treat the concrete with Temprid FX. You want to create a barrier around the bag of about two feet (60 cm) in each direction. That way, any cockroaches that escape from the bag will have no choice but to cross this barrier. If you cannot obtain Temprid FX, then purchase Talstar P. If you use Talstar P, mix one ounce (30 ml) to one gallon (3.8 liters) of water.

Remove the appliance from the bag and thoroughly inspect it for fecal matter, dead roaches, body parts, and of course, live roaches. If you see any of these sorts of activity, then repeat the treatment procedure and re-bag the appliance with a fresh IGR disk. Try opening it again in another 100 days. If everything looks good, then go ahead and bring it home. Otherwise, keep treating.

5. Electronics: Electronic items can be expensive, but if possible, you should still throw them away. However, if you must keep them, then follow these procedures.

Clean the item the best way you can without harming it. You want to make sure that there is no fecal matter on it anywhere.

Spray or wipe Nibor-D on any surface that will not be damaged by water.

Apply a bait on the bottom and back of the item.

Place the item in a sealed trash bag, along with a Gentrol IGR disk. Just like with small appliances, do not open the bag for at least 100 days. When it is time to open the bag, you should always do it outside. If possible, open the bag on top of concrete, like a driveway or sidewalk. Prior to opening the bag, make sure you treat the concrete with Temprid FX. You want to create a barrier around the bag of about two feet in each direction. That way, any cockroaches that escape from the bag will have to cross this barrier. If you cannot obtain Temprid FX, then purchase Talstar P. If you use Talstar P, mix one ounce (30 ml) to one gallon (3.8 liters) of water.

Remove the item from the bag and thoroughly inspect it for fecal matter, dead roaches, body parts, and of course, live roaches. If you see any of these sorts of activity, then repeat the treatment procedure and re-bag the item with a fresh IGR disk. Try opening it again in another 100 days. If everything looks good, then go ahead and bring it home. Otherwise, keep treating.

6. Books and other small items that do not contain metal: I think you get the picture now about throwing stuff away, so let's move right into how to treat these items. Any small items that do not contain metal can be microwaved.

Place the items in a plastic bag and seal it tightly.

Place the bag in a microwave along with ½ cup of water in a standard coffee cup. Microwaves heat liquids, so you must have half a cup of water in the microwave, or you risk damaging it. The steam that comes off the water will not harm the items in the bag because they are sealed in plastic.

I said that microwaves heat liquids. Well, guess what also contains liquids. That's right. Cockroaches contain liquids. So, when you microwave the water and the books, you're also going to microwave the cockroaches, and they will literally boil to death from the moisture in their bodies. Just make sure that I am not one of the roaches in the books.

I recommend microwaving on the highest setting for two or three minutes. You want to make sure that the water in the cup is boiling. If you see the bag start to expand, then shut off the microwave because you may be damaging the books. The roaches are most likely dead already because it is steam causing the bag to expand. That means the roaches are most likely cooked already.

Be careful when removing the cup from the microwave because it is going to be very hot.

Remove the items from the bag and check for activity. If everything looks good, then immediately place the items in a fresh bag and transport them that way to your new home. When it comes to German cockroaches, take no chances. Don't leave these items unbagged and unattended.

7. Clothes: Anything that cannot be washed in a washing machine and dried on high heat or cannot be dry cleaned, should be thrown away.

Clothes that need to be dry cleaned can be taken to a professional cleaner. Do your best to ensure that there are no roaches or eggs on the clothes. Place the clothes in a trash bag and seal it tightly.

Washable clothes should be placed in a sealed plastic bag and taken to a laundromat. Wash them in the hottest water possible and then dry them on the highest heat available until they are dry and hot. You know you did a good job when they are too hot to touch.

Some clothes can be microwaved the same way as books if they do not contain any sort of metal.

8. Shoes: All shoes, including the ones that you are wearing, should be sprayed with CB-80 Aerosol (See Figure 11) and placed in a sealed plastic bag for 24 hours. Open the bag outside to make sure that there are no cockroaches hanging out after treatment. You can pretreat a couple of pairs of shoes prior to the move date.

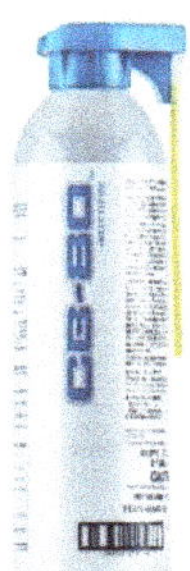

Figure 11 – CB-80 Aerosol

9. Wooden or unupholstered furniture: Yeah, yeah, throw it away if you can. You got the point now. These items can be treated fairly quickly, but you must do a thorough inspection and treatment to make sure that there are no hitchhikers.

Remove any drawers and inspect all surfaces for roaches or egg sacs. If you see any live roaches, then spray them with Shockwave 1. Egg sacs should be removed and microwaved. Simply spraying an egg sac with pesticide will not prevent the roaches from hatching.

Clean everything really well and make sure there is no fecal matter anywhere.

Spray all surfaces with Nibor-D that will not be harmed by water. I mean everything. Inside and out.

You can now bring this item directly to your new living area, but make sure that wherever you place it is surrounded by a treatment of Nibor-D. You want to do this as an insurance policy, in case you miss an egg sac or live cockroach. This is a good policy to adopt for all items you bring into your new home from the old one. It never hurts to create a barrier of Nibor-D around every item that you bring in from your old residence. Remember, it only takes one egg sac or one female cockroach to start a whole new infestation, so do not take any chances.

10. The moving truck: Make sure that the bed and the walls of the moving truck are treated with Temprid FX before you start the move. If you cannot obtain Temprid FX, then purchase Talstar P. If you use Talstar P, mix one ounce (30 ml) to one gallon (3.8 liters) of water.

Prior to entering the moving truck, make sure that you change into clothes that are not carrying cockroaches. This includes your shoes. Bag the clothes you wear and wash them later at a laundromat. If you do not have pretreated shoes, lightly spray your shoes with CB-80 before you start getting changed, so that they will be dry again when you put them back on.

11. Miscellaneous items like suitcases: Any items that I have not covered can be treated by one of the methods above.

By the way, did I mention that you should throw away any items that you do not need? I would like you to watch a very short video that was made by Animal Planet. It is less than 4 minutes long and does a great job of explaining the problem. I think after you see it, you will understand why you need to take all of these precautions. If you cannot take all of these precautions, then do as much as you can. Keep in mind that it just takes one female or one egg sac in any item you bring with you to start another infestation in your new home.

Animal Planet Video:
https://www.youtube.com/watch?v=RU9oOCYPros

14 HOW TO TREAT CARS

Cars are really hard to treat. The first thing you should do is make sure that you do not park in a garage. Parking in a garage will allow the cockroaches an opportunity to infest the house.

Step-by-step procedure:

1. Naturally, you want to start by thoroughly cleaning the car. Make sure that any food wrappers, crumbs, and so forth are removed. That means you need to vacuum really well. You should have a strict rule that nobody ever eats in your car, including you. Of course, if you drop food in the car, then you can make an exception for me. I am more than happy to eat in your car.

2. Next, you want to spray a mixture of Nibor-D and NyGuard IGR everywhere it will not be noticed and where it will not damage anything. You can spray it under the seats, in different places under the hood, and the entire trunk. Mix the Nibor-D 1/3 scoop to 16 ounces (473 ml) of cold water. Add 1.5 milliliters of NyGuard IGR. Measuring small amounts of NyGuard is easier if you purchase a measuring syringe.

3. Place a Gentrol IGR disc under one of the seats, one in the trunk, and one someplace under the hood. Replace them every 90 days.

4. Apply a pea sized dot of gel bait everywhere it will not be seen. Just use small dots and space them about 12 inches (30 cm) apart and use a different formulation of bait every 90 days. Domyown.com has a good selection of gel baits. Try to get four different baits, with different active ingredients. There is a list of baits at the end of this book and at www.guyspestsolutions.com.

5. In the summer months, always park your car in the sun, with all the windows shut because we do not survive well in high temperatures. You can also put an electric heater in the car to boost the temperature even more. If you can get the temperature over 120 degrees Fahrenheit (49 degrees Celsius), for several hours, that may do the trick. If you live in a cold climate, then you always want to keep the car outside because we do not survive well in the cold either.

6. German cockroaches are hitchhikers, so be careful that you do not bring us into the homes of friends and family members. We will get into anything you have in the car, and we will attach ourselves to your clothing.

15 HOW TO TREAT RESTAURANTS

Treating a restaurant is a little different than treating a house. The same principles apply, but you need to limit your use of Nibor-D a lot more.

Step-by-step procedure:

1. The first thing you need to do is apply permanent pesticides, like Nibor-D, boric acid, and diatomaceous earth (DE) everywhere a cockroach can possibly place its feet, that is consistent with label recommendations.

You need to be careful in restaurants because the label only allows you to use Nibor-D as a crack and crevice treatment or in inaccessible areas. It is somewhat unclear what constitutes an inaccessible area, but I take that to mean areas that would not be routinely touched by people. For example, my reading of the label would suggest that it is okay to apply Nibor-D under counters, under sinks, plumbing under sinks, on walls that are under counters, under and around appliances, or anywhere people would not touch on a regular basis.

Based on this interpretation, it seems clear that you are probably not supposed to mop all the floors with Nibor-D. Still, I do not see the harm in doing it. If the object of the restriction is to prevent food or utensils from touching the product, then I don't see a problem with treating the floors. If food or a utensil is dropped on the floor, the staff are supposed to throw that food away and wash any utensils. Still, the label does not seem to allow it. However, I would regard the baseboards as inaccessible. I mean, you humans do not routinely touch baseboards. That is more our domain. So, I would treat them. Also, I don't think humans routinely touch areas on the floor that are 6 inches (15 cm) along the walls, so I would create a band of Nibor-D on the floors around all the walls. Keep in mind that my interpretation of the label may not be consistent with the manufacturer's intentions, so use your own judgment.

Nibor-D can make the floors sticky. It can also make them look shiny or hazy. How it makes them look depends on the type of floor involved.

The point is that this is probably not something you want to do in the dining area. Therefore, you need to limit your Nibor-D floor treatment to the baseboards in areas that customers frequent. Again, my reading of the label is that it is okay to treat baseboards in dining areas.

Baseboards aside, you do not want to use Nibor-D to treat any surfaces that customers are going to see or may touch. So, I think it is okay to treat the underside of tables and chairs, but you do not want to treat the legs. Do not treat areas under tables that people are likely to touch. So, leave at least a 6-inch (15 cm) band around the perimeter of the underside of the tables.

2. You also want to make sure that you place Gentrol IGR discs under appliances and distribute them every five linear feet (1.5 meters) in the kitchen areas. Use double-faced tape to attach Gentrol IGR discs to the underside of tables and remember to replace them every 90 days for at least one year. Please refer to Chapters 8 and 9 to refresh your memory about the proper way to make these treatments.

3. I recommend applying boric acid dust or diatomaceous earth (DE) inside hollow table and chair legs. If you cannot find an access point to puff in the dust, then drill a small hole in an inconspicuous place.

4. If you use any sort of rolling carts, be sure to treat any hollow tubing with boric acid or DE, even if you need to drill a hole. Attach an IGR disk to the bottom of the cart with double-faced tape.

5. If your restaurant shares walls with other businesses or residences, then those areas are most likely also infested. Therefore, you will always have roaches continuing to enter your restaurant. However, if you do a good job applying the Nibor-D, boric acid, and DE, then any roaches that enter will be dead within a few days.

16 HIGH-RISE AND COMMERCIAL BUILDINGS

Regardless of the type of high-rise or commercial building involved, the treatment principles remain the same. So, this will apply to office buildings, hotels, apartment buildings, hospitals, and so forth.

Most of the treatment procedures that you need to know have already been explained in Chapters 8, 9, 12, and 15. The difference with high-rise buildings is that they have shafts or vertical chases that run from one floor to another. A good example of this would be elevators. Also, you will typically find waterpipe and electrical chases as well. These are used to run utilities from the basement to the roof. While these chases and shafts are both useful and necessary for high-rise buildings, they also provide an easy way for roaches to move from one floor to another. Therefore, these areas must be treated.

Step-by-step procedure:

1. Utility Chases: If there's an access door to a utility chase, then the easiest way to treat it is to blow boric acid or diatomaceous earth (DE) into the chase with an electric duster. Depending upon the height of the building, you may be able to do this from one access point near the roof. For very tall buildings, you may need to gain access from several locations. The idea is that you want the dust to come in contact with every area inside that chase. If there are no access panels available, then holes should be drilled to allow the duster tube to be inserted into the chase. You should always check with the facility manager to ensure that nothing in the chase will be damaged by the dust.

2. Trash Chutes: Trash chutes present a unique challenge because there's usually an access door on every floor and these chutes are often loaded with cockroaches. Unfortunately, the only practical way to treat a trash chute is with boric acid or DE. Since you do not want the dust to be blowing out of the access doors, all these doors must be securely closed and sealed with masking tape before the treatment is performed.

Out-of-service signs should be posted on all access doors during the treatment. If the trash cute is in a dedicated room, lock the room until the treatment is completed.

The dust can be introduced from the top floor and blown in with an electric duster. Enough dust should be applied to ensure that all surfaces inside the shaft are covered with the dust. Allow a couple of hours for the dust to settle and then remove the signs and tape from around the access doors.

Go floor to floor and treat all reachable surfaces in the shaft with Nibor-D, mixed with an insect growth regulator. Then attach an IGR disk to the inside of the chase door. Make sure you check them often and replace them if they are removed. As always, IGR disks should be replaced every 90 days.

3. Unmanned Dumb Waiters: Dumb waiters should be taken out of service permanently and then treated in the same manner as trash chutes.

4. Elevators: It is not a good idea to treat any sort of elevator with dust. Obviously, the movement of the elevator may disturb the dust, and you do not want to risk any dust getting into the elevator cars or finding its way past the elevator doors into the corridors. Therefore, elevator shafts must be treated with Nibor-D, mixed at 7.5 scoops per gallon (3.8 liters) of cold water. It is also a good idea to add an insect growth regulator. You should mix NyGuard IGR at a rate of 0.4 ounces (12 ml) of product to one gallon (3.8 liters) of mixed Nibor-D and water.

As always, the product can be applied with a tank sprayer. Since there is a lot of space to treat in an elevator shaft, it is recommended that a battery-operated sprayer be used for this purpose.

It is highly recommended that this treatment be done with the assistance of a qualified elevator mechanic. The best way to do it is to station the pest controller on the top of the car and then work floor by floor down through the elevator shaft to the basement. The job of the elevator mechanic is to operate the elevator and ensure the safety of the operation. Elevators must be taken out of service while this treatment is being performed.

5. Hospitals: Medical care buildings can present some unique challenges, but it is still possible to achieve total success.

When treating a hospital, keep in mind that we love food carts. Hospitals typically transfer food from the kitchen area to patient rooms on food carts. Therefore, these carts must be thoroughly cleaned, and all hollow areas must be treated with either boric acid or DE. If there is no access point to a hollow area, then a hole must be drilled to allow the treatment. The bottom side of the cart should be treated with Nibor-D. After the Nibor-D dries, attach an IGR disk to the bottom side of the cart with double-faced tape and replace it every 90 days.

Patient rooms, office areas, corridors, lobbies, waiting areas, radiology, the pharmacy, and that sort of thing should be treated the same way as a residence. Food service areas should be treated the same way as a restaurant.

6. Suspended Ceilings: The easiest way to treat suspended ceilings is by blowing boric acid dust or DE above the ceilings. This can be done with an electric duster.

Oftentimes commercial buildings will use ceiling voids as a chase to provide return air for the HVAC system. In this situation, dust should **NOT** be used. The top of the ceiling tiles must be treated with Nibor-D. It is also important to treat the top of the ceiling framework, the walls, and the underside of the floor above the ceiling with Nibor-D. The product should be mixed 7.5 scoops per gallon (3.8 liters) of cold water. It is also a good idea to add an insect growth regulator. You should mix NyGuard IGR 0.4 ounces (12 ml) of product with one gallon (3.8 liters) of mixed Nibor-D and water.

7. Nibor-D: In commercial buildings, it is always a good idea to limit the use of Nibor-D to areas that cannot be seen. That said, every area that cannot be seen, and that cannot be damaged by water, should be treated with Nibor-D. Boric acid or DE should be applied to all areas where it will not be disturbed by people, pets, or air currents.

8. Apartment Buildings: If you are a landlord who wishes to treat an apartment building, then it is important to gain tenant cooperation. You need to ensure that the tenants reapply Nibor-D to floors and other surfaces after cleaning. To do this, you must provide the Nibor-D for free, and it is important to offer an incentive to encourage compliance. A good way to gain tenant cooperation is to offer a 10% rent rebate.

The way it works is that you will enter the apartment every 90 days to change out the IGR disks, reapply bait, and inspect for compliance. If everything is clean and the Nibor-D is in place everywhere, then the tenant will receive a rebate check for 10% of the rent. I know this is expensive, but if you have compliance, then it will be over in a year.

Make sure there is a clause in the lease that will allow you to enter the apartment every 90 days, even without tenant consent. There should also be a clause that will allow the required access to drill the walls, apply boric acid into the wall voids, and perform all other treatments, as deemed necessary by the landlord. The lease should also contain a clause that provides for penalties for tenants who do not comply with pest control treatment requirements. For example, a 10% penalty on the rent may be imposed for those tenants who are found to be noncompliant.

17 MAKING AN ELECTRIC DUSTER

Electric dusters are a great way to blow dust into wall voids. Professional electric dusters are expensive, so I recommend that you make one yourself out of a cheap air pump that is made for inflating air mattresses and pool toys, and an empty peanut butter jar.

Always wear a respirator when applying dust like this. You are always going to get dust coming out of the voids you treat, and some may also leak out of the duster. It is also a good idea to wear rubber gloves and eye protection.

Step-by-step procedure:

1. To make one, get any empty jar that has a plastic, screw-on lid. Any size will work, but it's best to use one that you can hold easily. Drill two holes in the lid that are 3/8 inches or 9.5 millimeters in diameter. The best way to do that, to avoid damaging the lid, is to purchase a step drill bit. See Figure 12. Wrap a small piece of duct tape around it… so that you will know when to stop drilling.

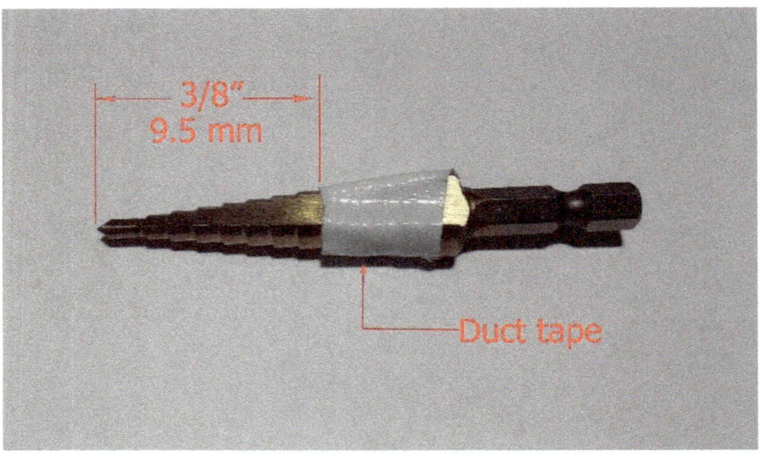

Figure 12 – Step drill bit

2. Make a mark about 3/4 of an inch (19 millimeters) from the side of the cap on opposite sides of the cap. It doesn't need to be exactly perfect. See figure 13.

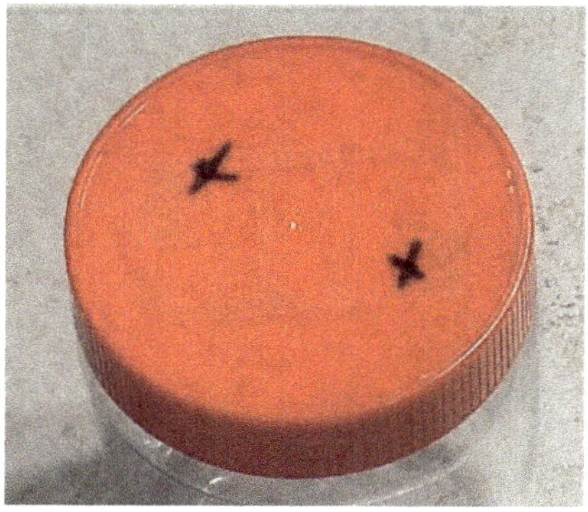

Figure 13 – Location of holes in jar cap

3. Now, drill a hole on both sides of the cap down to where you placed the duct tape on the step drill. See Figure 14.

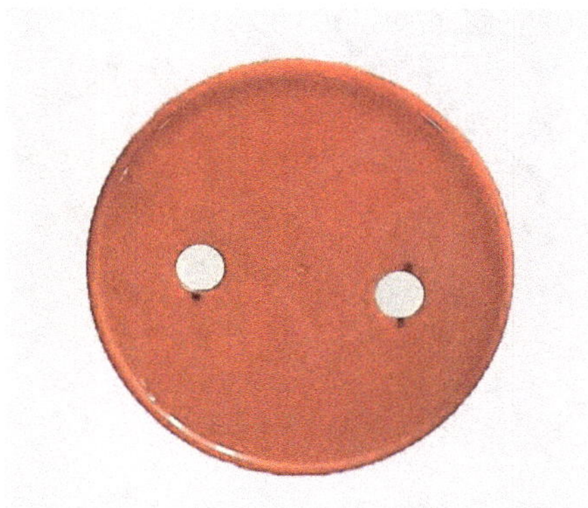

Figure 14 – Drilled holes in jar cap

4. Next, you need to get a straight, rigid object that you will use to straighten the longer hose that goes inside the jar.

In this case, I just used an old pencil that I had lying around, but any straight, ridged object will do.

Put the pencil into the jar and make a mark with sharpie that is even with the top of the jar. See Figure 15.

Figure 15 – Marking pencil at the top of the jar

5. Now, get a ruler and make a mark about 1/2 inch, or about 12 millimeters, down from the mark that you just made. See figure 16.

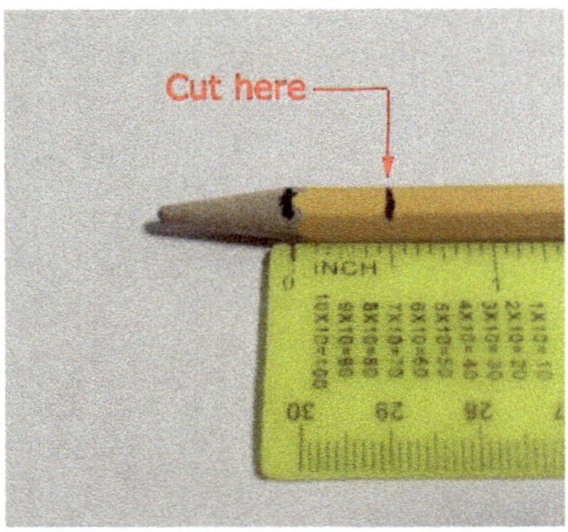

Figure 16 – Marking pencil cut point

6. Cut the pencil at the second mark that you made. I was able to do this with scissors. Well, Guy helped a little. I mean, I'm only a tiny roach, so the scissors are many times larger than I am.

7. You're going to need some clear plastic tubing that has an outside diameter of 3/8 of an inch, or about 9.5 millimeters. The inside diameter is not that important, but it will probably be 1/4 of an inch.

8. Make a mark on the tubing that is exactly the length of the pencil. See figure 17.

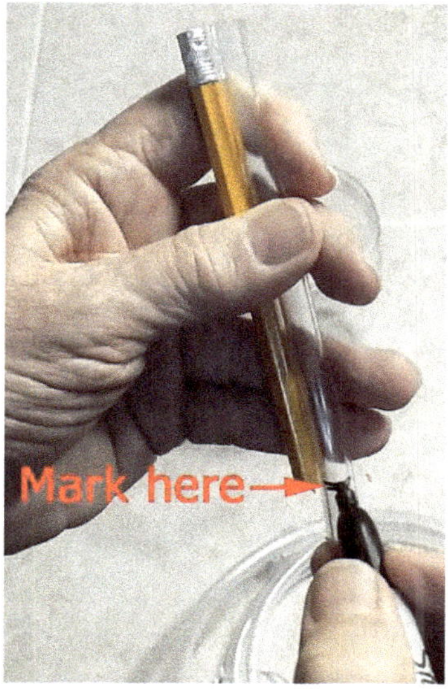

Figure 17 – Marking tubing

9. Get some duct tape and cut a piece that is about 3 inches, or about 8 centimeters long. I'm using white duct tape, but the color doesn't matter.

10. Next, you're going to need a small piece of waxed paper and a piece of cardboard. Put the wax paper over the cardboard to prevent the duct tape from losing its adhesive ability while you are cutting it.

11. Place the tape on top of the wax paper and use a sharpie to mark the center of the tape. See figure 18. Now cut the tape in half lengthwise with a utility knife. The edge of the tape that you did not cut is the factory edge.

12. Peel off the tape and you are ready for the next step. Do not discard the extra piece of tape because we will be using it later. In fact, you will need to make several of these short pieces.

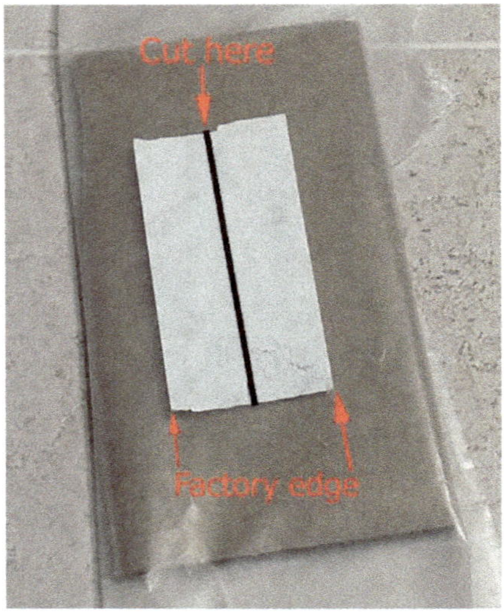

Figure 18 – Marking duct tape

13. Locate the mark on the tubing where you marked the length of the pencil. Measure about 3 inches (about 8 cm) from that mark and cut the tubing. The length is nominal. Make sure that you know the side of the tubing that is the same length as the pencil. See figure 19.

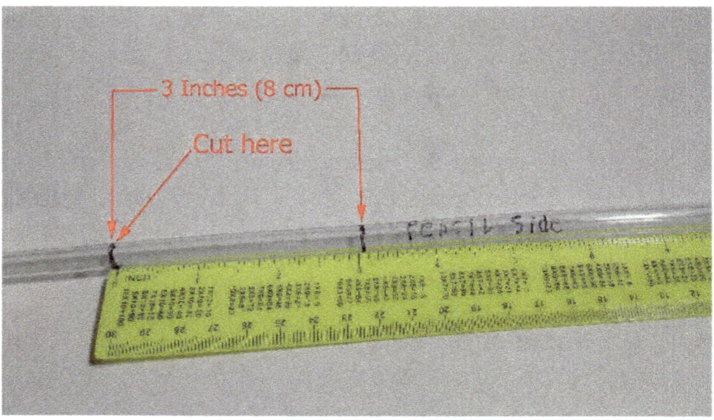

Figure 19 – Measuring tubing

14. Now, take a piece of the tape that you've cut and locate the factory edge. You need to use this side because you want the edge of the tape to be perfectly even when you wrap it around the tubing.

15. Carefully wrap the tape around the tubing right where the line is. You want the **UNEVEN** side of the tape facing the side of the tubing that is the same length as the pencil. See Figure 20. After you finish wrapping the first piece of tape around the tube, then wrap three more pieces over the top of it, again being careful so that the factory edge is perfectly even. We are using short pieces of tape because it makes wrapping the tape easier. The uneven side of the tape does not need to be perfect.

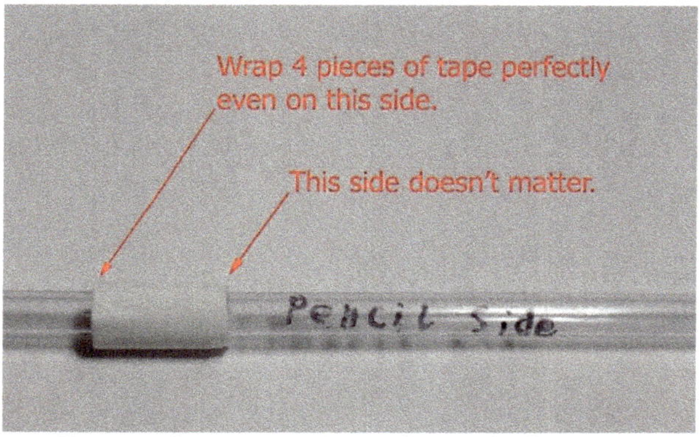

Figure 20 – Wrapping tube with duct tape

16. Attach the tube to the pencil with a piece of duct tape. You want the pencil to be about 1/4 of an inch, or about 6 millimeters, below the even edge of the tape. Use another piece of duct tape to attach the other end of the pencil to the tube. This should keep the tube reasonably straight. It doesn't need to be perfectly straight. You just don't want the hose bending a lot in the jar. See Figure 21.

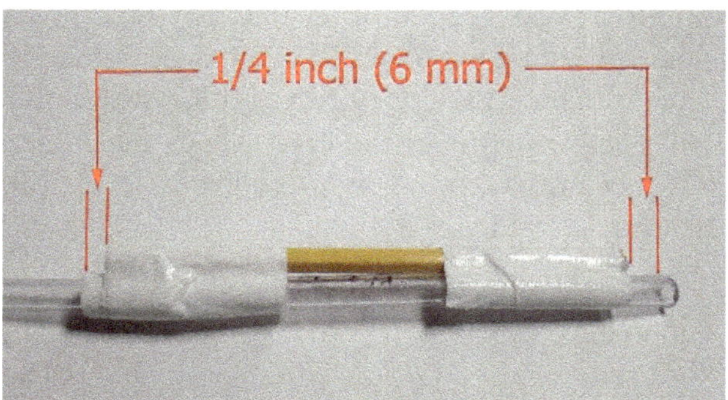

Figure 21 – Attaching the tube to the pencil

17. Insert the tubing from the bottom of the cap through one of the holes in the cap. It doesn't matter which hole you use. You want the pencil to extend into the jar when the lid is screwed on.

18. Use another narrow piece of duct tape that you cut and locate the factory edge. Wrap the tape around the tubing where it extends above the top of the cap, with the factory edge touching the top of the cap. Wrap the tape about 1/8 of an inch or about 4 millimeters too long, so that you know it's touching the cap. See figures 22 and 23. When you are wrapping the tape, make sure that you are pulling on the tube, so that it is tight up against the cap. The idea is that you want to create a seal on both sides of the cap. A little dust is probably going to still escape, but not that much. This is another reason why you need to wear a respirator when applying dusts.

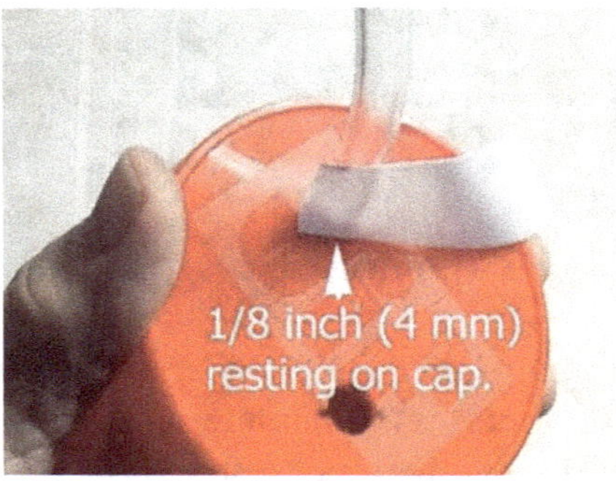

Figure 22 – Sealing tube to cap

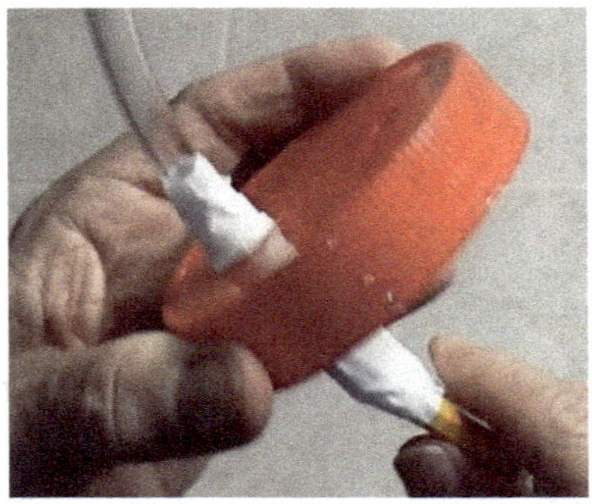
Figure 23 – Sealing tube to cap

19. Cut another piece of tubing that is about 12 inches long.

20. Just like before, you need another piece of tape that you cut, and you need to locate the factory edge. This time you want the **UNEVEN** edge of the tape touching the edge of the tubing where it was cut. The factory edge is going to contact the cap. Wrap the tubing so that the factory edge is nice and even. Again, just like before, you want to use four pieces of tape. See Figure 24.

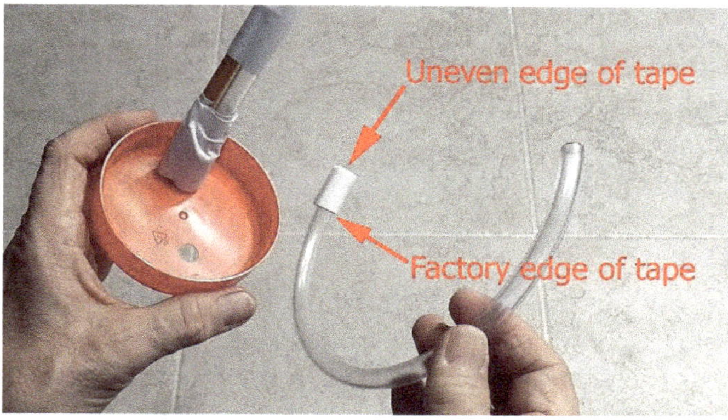

Figure 24 – Wrapping tape around exit tube

21. Now insert the tube from the bottom of the cap through the other hole in the cap. See Figure 25.

22. Just like with the other side, you want to make sure that you wrap a piece of tape tightly around the tubing where it comes out the top of the cap, to hold it in place to create a seal. See Figure 26.

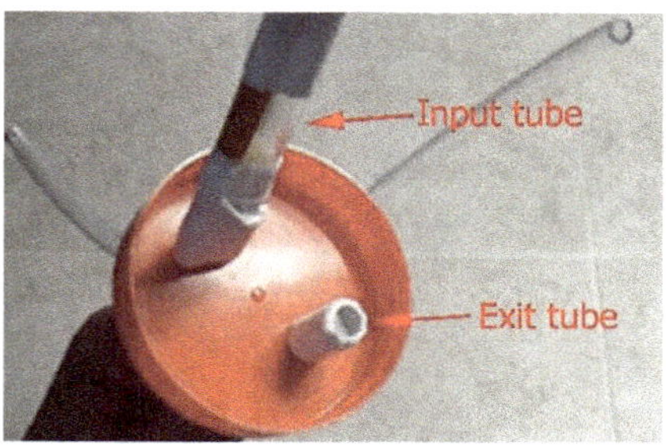

Figure 25 – Input and exit tubes

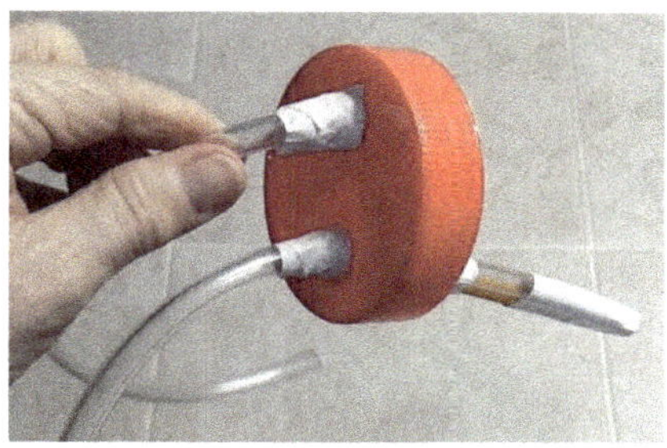

Figure 26 – Sealing tubes to cap

23. Now, just screw the cap onto the jar. The hose with the pencil is the input. The hose with the short side inside the jar and the long side outside the jar is the exit tube.

24. All you need now is an air pump that is suitable for inflating air mattresses or large pool toys. Guy uses this tiny air pump in Figure 27, but for even more power and more dust coming out of the duster, you can invest in a bigger pump.

25. Locate the reducer tip like the one in Figure 27. If it is attached to the other reducer tips, simply cut it off. If the end of the reducer tip is not a straight opening, cut off the end so that the air will shoot straight out. Now attach the reducer tip to the air pump and insert the tip into the inlet hose on the duster. That would be the hose that leads to the pencil.

26. When you put the dust in the jar, make sure there are no lumps in it. You don't want lumps getting caught up in the tubing. The clear jar makes it easy to see lumps, and you can break them up by stirring the dust with a stick or a screwdriver. After you have all the lumps out, screw the jar onto the duster.

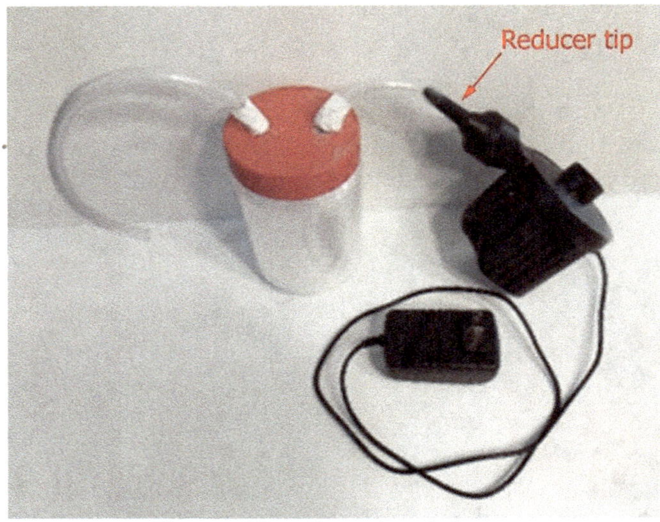

Figure 27 – Completed duster

27. This is very important! Do not fill the duster jar more than halfway.

The duster puts out a lot of dust, but it works even better if you lightly shake the jar while the duster is operating. This will cause a massive amount of dust to be released at the end of the duster tube.

18 FREQUENTLY ASKED QUESTIONS

Q. How long should I wait between Phase 1 and Phase 2?

A. There is no hard, fast rule on this. The purpose of Phase 1 is to achieve a quick knockdown of the population and to permanently treat cabinets and other surfaces. You should reach the maximum effect in about 30 days, but you can do Phase 2 sooner than that if you like. Do not wait longer than 90 days to do Phase 2 over the Phase 1 areas. Remember that Phase 1 is usually only done in the kitchen and bathrooms, so you can start Phase 2 in all other areas of the home, just as soon as Phase 1 is completed in the kitchen and bathrooms. By the time you get finished with Phase 2 in the rest of the house, 30 days have probably gone by already, so you can start on Phase 2 over the Phase 1 areas. If you finish Phase 2 early in the rest of the house, then you can go over the Phase 1 areas just as soon as the population is reduced to an acceptable level. So, it's kind of a judgment call on your part.

Q. I have a minor infestation, and I don't see a lot of roaches. Can I just skip Phase 1 since it is for a quick knockdown?

A. No. While phase one is designed for a quick knockdown, you are using Nibor-D, which is a permanent pesticide that must be applied, even if you are not seeing many roaches. You also must have the Gentrol Point Source IGR discs in place as well. You could skip using the bait, but it's not a good idea. However, if you are not seeing many roaches, then you could skip the Temprid FX altogether and just use Nibor-D. Instead of using D-Fense dust, you can use boric acid or diatomaceous earth in cracks and crevices where you need to apply dust.

Q. What is the difference between Nibor-D and boric acid? Can I use them interchangeably?

A. Nibor-D and boric acid are very similar products. Both can be used as a dust, but only Nibor-D can be mixed with water. Manufacturers claim that you can mix boric acid with water, but I have found that not to be the case. Nibor-D is a better choice for wet applications. Boric acid is a bit less expensive than Nibor-D and it works fine for dry applications. So, while you can use Nibor-D as a dust instead of boric acid, you cannot use boric acid for liquid applications.

Q. What is the difference between boric acid and diatomaceous earth?

A. Boric acid is derived from boron, which is a naturally occurring element in the Earth's crust. Diatomaceous earth is made from the fossilized remains of tiny aquatic organisms called diatoms. While these products are different in terms of origin, they work pretty much the same way. We hate to get anything on our feet. When we cross either of these materials, it sticks to our feet, and we ingest it to get it off. It then cuts holes in our stomachs, and we dehydrate to death. It works the same way if you make a bait from either of these materials. If we eat the bait, we will die from dehydration. If we get it on our exoskeleton, it will also cause us to dehydrate to death. Diatomaceous earth is less expensive and has the advantage that it can be mixed with water. The disadvantage of diatomaceous earth is that Guy has not tested it in field conditions, so he cannot guarantee the same results as using boric acid.

Q. I am confused about the dilution rates of Nibor-D. Sometimes you say 2.5 scoops per gallon and sometimes you say 7.5 scoops per gallon. Also, when you explain mixing smaller amounts, the dilution rate is not exactly the same as when mixing a gallon. Can you clear that up for me?

A. Always use 2.5 scoops of Nibor-D per gallon of cold water for mopping floors and treating carpets. Always use 7.5 scoops of Nibor-D per gallon of cold water for all other applications. Since it is hard to measure partial scoops of Nibor-D, you cannot get the measurement exactly right. That's okay because you can mix the product anywhere from 2.5 scoops per gallon to 7.5 scoops per gallon, so close is good enough when mixing amounts smaller than one gallon.

Q. Why do I need the Gentrol Point Source IGR Disks and how do they work? Why do I need to replace them every 90 days?

A. The IGR disks work even slower than borates. You must replace them every 90 days for a year because they don't kill roaches at all. They simply stop us from breeding. German roaches have about a 100-day life cycle, and it can take up to 3 life cycles for us to be totally eliminated. Unfortunately, the only way to treat appliances and electronics is with IGR disks, and the disks only last 90 days. So, you need to see this as a one-year treatment plan. If you do a good job with the Nibor-D, and you apply it on all surfaces under and around appliances and electronics, then any roaches that try to move to another area of your home will die. If we remain where we are, we will not be able to breed and we will die out in about a year.

Q. You said that it is a good idea to mix an insect growth regulator with Nibor-D, but you are also recommending the use of Gentrol Point Source IGR disks. Why are you recommending that I use two different IGRs?

A. Using two different IGR products may initially be a bit of overkill. However, you must keep using an IGR for an entire year. The liquid IGRs will not last that long and it is time-consuming to reapply them. It's a lot faster and easier to just replace the IGR disks. It does no harm to use two IGRs, and they are not exactly formulated the same way, so one may be more effective than the other.

Q. I completed both Phase 1 and Phase 2, but I am still seeing roaches. What's going on with that? Did I do something wrong?

A. Phase 1 will knock out a lot of roaches, and you will see the population go down fast, but there are lots more of us that you are not seeing. Phase 2 will kill those roaches, but you may still see some of us coming out over the next few months. There will be less and less of us until we are totally wiped out. If you are still seeing roaches after about 6 months, then you may have missed a void somewhere. This is very easy to do. Even the best pest controllers can miss a void. If this happens, simply place IGR disks near where you are seeing the activity and replace them every 90 days for a year. Keep in mind that, if you did a good job with the Nibor-D, those roaches are going to die when they cross that material. So, you do have us isolated now. The IGR disks should finish us off within a year.

Q. I have not seen any roaches for a couple of months. Can I just stop the treatment early?

A. The biggest mistake that people make, even the pros, is they underestimate German cockroaches, and that is the main reason why they fail. You may have reduced the population by 99%, but if even one female roach survives, or just one egg sac, then the infestation can start up again if you remove the Phase 2 protections. This is especially true for the IGR disks. It is very important to replace the disks every 90 days for an entire year.

Q. Can exposing German roaches to cold temperatures kill them?

A. We do not like cold weather, and we cannot breed when the temperature is below about 40° Fahrenheit (4.44 degrees Celsius), so for example, if a car sits in that temperature for a really long time, then that can end the infestation. Unfortunately, we are talking about months, and it is likely that spring will arrive before all of us are dead. It only takes one egg sac to survive, and you will be off to the races again. For cockroaches to actually die from the cold, the temperature would probably need to be below zero degrees Fahrenheit (-17.8 degrees Celsius) for at least several days. Exactly how long this temperature would need to be maintained is unclear, but the colder the better, and the longer the better. This will probably work best in very cold climates like Maine, Vermont, North and South Dakota, Alaska, and places like that, where you often see temperatures drop below zero for an extended period. It is possible if a car is sitting outside for a long weekend when the temperature is at or below something like -20°F (-6.67°C), that the roaches are not going to survive. So, while parking your car outside is always a good idea, it is probably not an effective way to eliminate roaches in most locations. It certainly will help, but you're also going to need to use pesticides as well. You may be able to freeze us to death in homes as well in those very cold climates, but you must drain all water pipes first.

Q. Can German cockroaches survive high temperatures? I was told that I could put really good heaters in the house and get the temperature high enough to eliminate the roaches. Is this true?

A. The answer to that question is "maybe." German cockroaches will not survive for long if they are exposed to temperatures above 130 degrees Fahrenheit (54.44 degrees Celsius). In fact, death may occur within minutes at that temperature. Sounds good. Right? Well, not so fast. We are usually living in the walls, so you need to get the inside of the wall voids to reach the desired temperature. While it is somewhat easy to get the inside air in the house above 130°F, the walls are a bit more difficult. Drywall is an insulator, so it needs to get really hot in the house before the temperature in the walls gets hot enough to kill the roaches and the egg sacs. So, while it is possible for this treatment to work, it is not a sure thing. All it takes is for one egg sac to survive and you have another infestation. This is why I do not recommend heat treatments to eliminate a German cockroach infestation.

Q. I occasionally see albino roaches that are white. Is this unusual?

A. This is not unusual. But these are not albino roaches. The reason some German roaches appear white is because they have recently molted. Molting is a process where cockroaches shed their old exoskeleton to grow a new one. When this happens, they lose most of their pigmentation, and that makes them appear white for a short period of time. They will soon regain their natural color.

Q. I live in a mobile home. How do I treat voids in paneled walls and ceilings? If I drill through them, they cannot be repaired.

A. Most of the time you can find trim pieces that can be removed that will allow you to drill a hole that can be covered up when you replace the trim.

You may find trim pieces near the ceiling. However, if you cannot find trim pieces, you may be able to remove the baseboards. Mark the wall at the top of the baseboard before removing it. Drill the hole just below the mark. The baseboard will cover the hole when you reinstall it. Make sure that you are drilling into the void and not a solid piece of wood. Applying the boric acid or diatomaceous earth at the bottom of the void is not as effective, but it's the best you can do.

If you cannot find trim pieces that will allow ceiling treatment, then wipe Nibor-D, mixed with NyGuard IGR, onto and around all ceiling fixtures, such as lights, fans, HVAC registers, and so forth. After it dries, puff boric acid or diatomaceous earth in all cracks and crevices in the ceiling. The idea is that you want to force the roaches to cross this barrier when they try to leave the ceiling void. The NyGuard IGR may stop them from reproducing, thereby ending the infestation in the ceiling.

These alternative treatments may not be as effective as drilling into voids correctly, so you may need to keep Phase 1 and Phase 2 materials in place longer to ensure that roaches that leave untreated voids will die when they cross the Nibor-D or boric acid dust.

19 CONCLUSION

As you can see, eliminating a German cockroach infestation is not exactly an easy task. Please keep in mind that the longest journey begins with a first step. Try not to look at this in its entirety. Projects like this are much easier to accomplish if you break them down into smaller pieces. The most important thing is to get started. The sooner you get started; the sooner you will be rid of us. Also, keep in mind that the key to success is thoroughness. If you want us out of your life, then you must leave no stone unturned.

I would love to know how successful you end up being after completing this process. Unfortunately, German roaches only get to live about 100 days, and I have already been around for a while. Therefore, I think it is a reasonable conclusion that I will probably be dead before you even buy this book. Perhaps a lot sooner than that if Guy doesn't release me from the jar he keeps me in. Trust me … it's not easy being a roach.

Be well my human friends and enjoy a roach-free home.

ACKNOWLEDGMENTS

I would like to thank all of my family and friends who assured me that I was indeed capable of writing a book. Apparently, they were all correct, because, well, here it is. I want to especially thank my wife, daughter, and brother, who assisted me with copy editing. I could not have done it without them.

I would also like to thank the many, many viewers of my YouTube channel who encouraged me to write this book. Many of them constantly "bugged" me for over two years to do it. Had it not been for them, I probably would not have done it.

Finally, for all of you reading this, I would like to thank you as well. You have made writing this book worthwhile. I am so sorry if you have a cockroach infestation, but after you eliminate these guys, I wish you a totally cockroach-free life going forward.

Be well my friends.

Guy

ABOUT THE AUTHOR

Guy Amato has been involved with the pest control industry since 1981. His initial training took place on Long Island, New York and was conducted by Dr. Austin Frishman Ph.D., a former professor of entomology at the State University at New York at Farmingdale. Guy has done significant work for the US Federal Government, mostly with German cockroaches. He has performed pest control services in the Midwest, Eastern, and Southern United States. Guy's business experience includes owning three corporations. Guy is currently retired and now provides free pest control advice on his YouTube channel, www.youtube.com/@guyspestsolutions, and his website at www.guyspestsolutions.com. In addition to pest control, Guy has experience in engineering and the construction of homes. The procedures for eliminating German cockroaches that are depicted in this book were created and developed by Guy personally and are based on his extensive knowledge of German cockroaches, pest control products, and building construction techniques.

In the interest of full disclosure, Ronnie the Roach is a fictional character that was used for entertainment purposes. Seriously… roaches can't really write books.

MATERIALS LIST

Nibor-D - domyown.com, solutionsstores.com, Amazon

Gentrol Source Point IGR Disks - domyown.com, solutionsstores.com, Amazon

Temprid FX - domyown.com, solutionsstores.com, Amazon

D-Fense Dust - domyown.com, solutionsstores.com, Amazon

Boric acid - domyown.com, solutionsstores.com, Amazon

Diatomaceous Earth - domyown.com, solutionsstores.com, Amazon

Nuvan Directed Spray Aerosol: domyown.com, solutionsstores.com

NyGuard IGR - domyown.com, solutionsstores.com, Amazon

CB-80 Aerosol - domyown.com, solutionsstores.com, Amazon

Shockwave 1 - domyown.com, solutionsstores.com, Amazon

Handheld duster by TRIWISH – Amazon

Koblenz WD-2L Portable Wet-Dry Vacuum – Amazon

Franklin Sensors ProSensor 710+ Stud Finder - Amazon

Air pump - Amazon

Respirator - Amazon

Rubber gloves - Amazon

Eye protection – Amazon

Baits (Select any four of the baits below. They are available from domy-own.com):

Advion Cockroach Gel Bait

Vendetta Plus Cockroach Gel Bait

Vendetta Nitro Cockroach Gel Bait

Optigard Cockroach Gel Bait

Alpine Cockroach Gel Bait

GLOSSARY

Active Ingredient
The chemical component in a pesticide responsible for killing or affecting cockroaches.

Apartment Modification
Adjustments to treatment methods specifically designed for multi-unit dwellings to prevent reinfestation from neighboring units.

Bait
A pesticide formulation mixed with food that cockroaches consume and carry back to their hiding places, spreading the active ingredient.

Boric Acid
A natural, permanent dust pesticide that damages a cockroach's exoskeleton or damages a cockroach's digestive system after ingestion, leading to dehydration and death.

Cannibalism (Cockroaches)
The behavior in which cockroaches feed on dead or dying members of their own species, usually when food is scarce or populations are overcrowded.

Cockroach Infestation
The presence of reproducing cockroaches in a structure. Even a single sighting usually indicates a large hidden population.

Crack and Crevice Treatment
Application of a pesticide into small openings where cockroaches hide or travel.

Diatomaceous Earth (DE)
A natural, permanent dust pesticide that damages a cockroach's exoskeleton or damages a cockroach's digestive system after ingestion, leading to dehydration and death.

Dust (Pesticide Form)
A dry, powdered formulation applied into cracks, crevices, voids, walls, and hidden areas.

Fast-Kill Pesticide
Usually a non-repellent pesticide that kills cockroaches within hours, allowing them to spread the poison before dying.

Fogging
A method of dispersing pesticide as a fine mist. Produces quick visible results but is generally ineffective for elimination and may worsen infestations.

German Cockroach (Blattella germanica)
A highly adaptive indoor cockroach species known for rapid reproduction and resistance to pesticides.

Hitchhiking (Infestation Spread)
The process by which cockroaches travel via belongings, clothing, or items from infested locations to new environments.

IGR (Insect Growth Regulator)
A substance that prevents cockroaches from reproducing, effectively acting as birth control for the population.

Infestation
A condition where cockroaches are actively living, breeding, and spreading within a structure.

Instant-Kill Pesticide
Usually a repellent pesticide that kills immediately on contact.

Nibor-D
A borate-based pesticide used as both a liquid and dust treatment, providing long-lasting control.

Non-Repellent Pesticide
A pesticide formulated so insects cannot detect it, allowing them to walk through the treated area without avoidance.

Nymph
A juvenile cockroach that has not yet reached adulthood.

Ootheca (Egg Sac)
A capsule produced by female cockroaches containing multiple eggs (typically 30–40).

Permanent Pesticide
A pesticide that remains effective indefinitely if undisturbed (e.g., boric acid and diatomaceous earth).

Phase 1 (Quick Knockdown)
The first stage of treatment focused on rapidly reducing the cockroach population using a combination of fast acting pesticides and permanent pesticides that are designed to last forever.

Phase 2 (Finish Us Off)
The second stage focused on eliminating hidden populations by treating wall voids, furniture, and inaccessible areas.

Repellent Pesticide
A pesticide that cockroaches can detect and may avoid.

Resistance (Pesticide Resistance)
The ability of cockroaches to survive exposure to a pesticide and pass that resistance to future generations.

Residual (Pesticide Residual)
The length of time a pesticide remains effective after application.

Slow-Kill Pesticide
A pesticide that kills over several days, allowing for distribution to other cockroaches.

Spiracles
Openings in a cockroach's body used for breathing; can be blocked by substances like soapy water.

Temprid FX
A professional-grade pesticide used for baseboard and surface treatments.

Tracheae
Internal tubes that carry oxygen throughout a cockroach's body.

Wall Void
The empty space between wall studs where cockroaches commonly hide and breed.

INDEX

www.ingramcontent.com/pod-product-compliance
Lightning Source LLC
LaVergne TN
LVHW022305200926
848481LV00004B/1228
9798950722004